Fly-by-Wire

A Historical and Design Perspective

Other SAE books on this topic:

Aircraft Flight Control Actuation System Design
by E. T. Raymond and C. C. Chenoweth
(Order No. R-123)

For information on these or other related books, contact SAE by phone at (724) 776-4970, fax (724) 776-0790, e-mail publications@sae.org, or the SAE web site at http://www.sae.org.

Fly-by-Wire
A Historical and Design Perspective

Vernon R. Schmitt

James W. Morris

Gavin D. Jenney

SAE INTERNATIONAL ®

Society of Automotive Engineers, Inc.
Warrendale, Pa.

Library of Congress Cataloging-in-Publication Data

Schmitt, Vernon R.
 Fly-by-wire : a historical and design perspective / Vernon R.
Schmitt, James W. Morris, Gavin D. Jenney.
   p.   cm.
 Includes bibliographical references and index.
 ISBN 0-7680-0218-4
 1. Fly-by-wire control.   I. Morris, James W.   II. Jenney, Gavin
D.   III. Title.
TL678.5.S35   1998   98-22456
629.135'5—dc21   CIP

Cover photos: F-16 (top), Boeing 777 (bottom).

Copyright © 1998 Society of Automotive Engineers, Inc.
         400 Commonwealth Drive
         Warrendale, PA 15096-0001 U.S.A.
         Phone: (724) 776-4841
         Fax: (724) 776-5760
         E-mail: publications@sae.org
         http://www.sae.org

ISBN 0-7680-0218-4

SAE Order No. R-225

Contents

Acknowledgments

The Wright Brothers would have been proud of what aircraft F-4 USAF #12200 demonstrated on April 29, 1972, at St. Louis International Airport. The problems the Wrights had had with mechanical controls were finally put to rest over 60 years after their first flight with the first aircraft to be completely fly-by-wire. It was, as McDonnell Douglas Corporation so aptly predicted, "F-4, Fly-By-Wire . . . Research Platform for the Future." For flight control systems, this event was analogous to that first voice to be transmitted and heard over a telephone wire.

The Air Force Flight Dynamics Laboratory provided resources and management personnel for this program and the McDonnell Douglas Aircraft Company, under contract, carried out the work of system installation and flight tests.

Acknowledgment is owed to the Air Force Flight Dynamics Laboratory staff who gave of their time and talents to make this book possible:

- Morris Ostgaard initiated the Douglas Long Beach Program, the first in a fairly long list of efforts either directly or indirectly related to fly-by-wire.

- James Morris, Program Manager of 680J Survivable Flight Control System Project from 1969 through 1974, did a superb job in directing this program to a successful conclusion. He disseminated throughout the Air Force and industry the fly-by-wire flight control systems technology. He also wrote the following sections of this book: Phase IID. Flight Demonstrations and Air Force Evaluation; Phase IIE. Precision Aircraft Control Technology and Reliability/Maintainability of SFCS Fly-By-Wire; and Technology Transitions and Application.

- Gavin Jenney was Director of Air Force In-House Facilities, including the B-47 Program. Gavin also designed the fly-by-wire system used on this program and was one of the flight test team. He is author of many Air Force technical documents on fly-by-wire and is currently President of Dynamic Controls, Inc. He reviewed and made numerous suggestions to improve this book.

- F/L Paul Sutherland, a Canadian Air Force exchange officer, came on the scene at a very opportune time. He took over Phase II of the B-47 In-House Program and Phase II of the Sperry Phoenix Effort. He organized and chaired the Government-Industry Symposium on Fly-By-Wire given at Wright-Patterson Air Force Base.

- Major General Robert R. Rankine, HQ AFSC, while serving as a Major during his early military career, was a Section Chief in the Control Elements Branch of the Flight Dynamics Laboratory. He offered early encouragement of this publication, as illustrated by this excerpt from a June 11, 1992, letter to HQ AFMC:

  Technology for fly-by-wire was developed by Wright Laboratory in the 1960s and first saw application in a production airplane in the F-16. Mr. Harry Hillaker, formerly of General Dynamics and now a member of the Air Force Scientific Advisory Board, has told General Yates that fly-by-wire in the F-16 would not have been possible without the Flight Dynamics Lab (now part of Wright Lab). . . . I believe a monograph on this subject would be of considerable value to the AFMC technical community, since it represents a USAF-led technology development that has had a significant impact on world aviation. . . . I believe such a monograph would be popular with a multitude of engineers throughout the aerospace industry who have been working fly-by-wire issues for decades.

- Ed Snyder and other personnel of the Air Force Materials Laboratory indoctrinated the McDonnell Douglas engineers on the survivability aspects of hydraulic fluid Mil-H-83282 over Mil-F-5606 (the then standard Air Force fluid), mainly the performance at higher temperature and the higher bulk modulus. The new fluid was used in A/C 12200 throughout the Fly-By-Wire Program and is now the standard hydraulic fluid in Air Force aircraft.

There are many other individuals who deserve to be mentioned as they played a part in establishing the technology we use. However, I feel we must set a limit, and sincerely hope they'll understand the situation.

I offer special thanks to my wife, Betty. With WordPerfect, helpful suggestions, encouragement, and marital support, she made this book possible.

Vernon R. Schmitt
Former AFFDL Project Engineer, Task 822503

Chapter 1

Introduction

Fly-by-wire in aircraft flight control design is more than adding a simple wire—it is a sophisticated system that changes the way aircraft are designed and the way they fly. Previous works on fly-by-wire have omitted or inadvertently bypassed many of the details and explanations of the system's design, choosing instead to highlight only its application. Our purpose then is to fill this void by describing the "how" and the "why." This book was prepared and written by people who directed or staffed the fly-by-wire research and development programs because, as time passes, without a formal account of the key events and activities of the initial R&D phases, knowledge of the pitfalls, problems, and the technology employed to overcome them may be lost.

The data presented herein are not intended as the complete documentation of proceedings and results, as the programs described herein cover a time period of approximately 12 years, from 1961 through 1973. (A full collection of data and applications of fly-by-wire technology in some form would span over 45 years.) Most of the R&D was performed under contract; work referred to as "in-house" was accomplished at Wright-Patterson Air Force Base (WPAFB) in Dayton, Ohio, by contractor and Air Force military and civilian personnel.

The Air Force spent millions and millions of dollars to promote and advance fly-by-wire technology to a level where it could be used as a practical design technique on new aircraft. Furthermore, had these programs not been undertaken and had they not been successful, fly-by-wire would not have been used on the design of flight control systems of military aircraft for at least for another 30 years—and perhaps not even then.

Unlike some design techniques that can be proven by mathematical models, computers, simulators, and perhaps laboratory models, fly-by-wire required the design, construction, tests, and flight tests of flightworthy hardware. With many R&D programs, the ratio of paper to hardware seems to average 75% paper to 25% hardware—on fly-by-wire the R&D ratio was more like 10% paper to 90% hardware, which added considerably to the cost factor.

The challenge for the designers of the fly-by-wire system was not only meeting the technical requirements, but also creating a system acceptable to pilots in terms of handling qualities. The pilots wanted the aircraft to fly like, or better than, the conventional designs they had been flying. In this respect, the tremendous design benefits and tools the fly-by-wire method provides are too readily overlooked. We define fly-by-wire as transmission of electrical signals from the pilot and as inputs to the flight control system. Fly-by-wire–designed aircraft offer the designer potential capabilities that were hardly possible before:

- Applied computational methods[1]—pilots compute how they want to fly while they are flying.
- Handling qualities that permit pilots an easy means to alter the "feel" of their controls.
- Redundancy that not only provides multiple signal paths but also voting schemes to sense a failure automatically and switch channels should one fail, without affecting control or flight of the aircraft.[2]

Other possibilities now exist which are not often mentioned: the ease of blending other systems with flight control[3] and the ability to check or test the system automatically on a "go-no-go" basis.

The development of actuators for fly-by-wire systems has been a continuous effort for the past 20 to 25 years. In 1995, the flight tests of what are called "smart actuators" showed real, positive results. The Boeing Company has applied many of these early potential advantages in the flight control design of their new Fly-By-Wire #777. (See Appendix C.)

---

1. During Phase II of the B-47 Program the C* concept was used; it was computer designed, installed, and flight tested with good results. On the F-16 the neutral stability point was moved and with movement of the leading edge slats (computed), we have *relaxed static stability.* Experimentally on the F-111 we created the MAW (mission adaptive wing), which permitted changing the wing camber in flight.

2. The all-important action for changing concepts into reality is *mechanization.* Mechanization was accomplished on the B-47 program. In fact, the Survivable Flight Control System mechanization on the 680J Program made fly-by-wire a reality.

3. The B-2 and F-117 both employ this part of fly-by-wire. Thus, when we start to blend signals that are computerized (or computed) we can blend the various systems: avionics, flight control, and so on, ad infinitum. In the future, we can adapt such techniques to automatic flight for take-off, cruising, and landing.

Chapter 2

Background of Fly-by-Wire

Definitions

What is meant by *fly-by-wire*? A simple, abbreviated description might be: fly-by-wire is a flight control design wherein a mechanical link is replaced by an electrical one. Obviously, this description must be expanded to comprise the link's purpose or function and the items being linked.

In some respects, a better understanding of fly-by-wire might be obtained if one were to visualize the pilot moving the controls. On the controls are attached electrical devices that generate and transmit electrical signals to another member that is remotely located in the aircraft. The controlled member moves in step with the pilot's commands. Further, when sensor and computed outputs are added to the pilot's commands, the pilot has the capability of controlling the aircraft motion about its referenced axes. Sensing components, such as gyros and accelerometers, along with computing devices and actuators are part of a fly-by-wire flight control system.

Over the years a number of definitions of fly-by-wire have been proposed. In 1967, this was the accepted definition: "A fly-by-wire flight control system is an electrical primary flight control system employing feedback such that the vehicle motion is the controlled parameter." Eventually, this definition was superseded. As of 1995, the one now accepted as standard and used by U.S. industry is as follows:

Fly-by-wire A flight control system wherein vehicle control input is transmitted completely by electrical means. (This definition applies to digital designs as well as analog.)

Other terms used with fly-by-wire systems are:

Flight control system A system that includes all aircraft subsystems and components used by the pilot or other sources to control one or more of the following: aircraft flight path, attitude, airspeed, aerodynamic configuration, ride, and/or structural modes.

Autopilot A portion of the aircraft's flight control system that automatically initiates flight corrections by sensing deviations from a fixed reference; the autopilot performs the necessary functions, including computations and actuation, to maintain the aircraft on a steady preset course and attitude without assistance from the pilot. The term "autopilot" is short for *automatic pilot.*

Pseudo fly-by-wire A fly-by-wire control system with a mechanical backup that is normally disengaged.

Redundancy. The existence of more than one means for accomplishing a given function.

Power-by-wire actuation An integrated servoactuator that incorporates an electric motor to receive power from the aircraft main electric power system in lieu of an actuator connected directly to the aircraft main hydraulic system.

Definitions for other frequently used terms can be found in documents listed in the References section.

History

World War II Autopilots

Fly-by-wire has been around in one form or another for at least 50 years. In 1943 the C-1 Autopilot was developed and installed in the B17E bomber aircraft for use during World War II. The primary purpose of this equipment was for bombing. It provided a stable platform for the bombardier to do the proper target sighting, and it eliminated undesirable motion of the aircraft at the time of the bomb release. The pilot could also make use of the autopilot, especially on long mission flights over water.

The C-1 Autopilot was electrical. There were potentiometer-type transducers on the vertical and directional gyros to sense changes in roll, pitch, and yaw. Voltages from these transducers were fed to electronic amplifiers, the output of which operated solenoids or servomotors which, in turn, had cable drives attached to the flight control surfaces (the elevator, rudder, and ailerons). The pilot would trim the aircraft for straight and level flight, then engage the autopilot. Thus, in a sense, he was flying on a fly-by-wire system. The C-1 Autopilot was built by the Honeywell Corporation. The E-4, used on other military aircraft, was built by the Sperry Corporation.

The A-12 Autopilot, also built by Sperry and used on operational aircraft in 1946, is representative of the type of equipment that was installed and flying on operational aircraft. Figures 1 and 2 give some basic data about its operation. Fig. 1 shows a cutaway view of the vertical gyro. Fig. 2 shows the electrical signal flow of this autopilot. The flight amplifiers shown in the figure are electronic types. The actuators are electromechanical, with a clutch and cable drive connected to the flight control surface.

The Guided Missile Era

After World War II, the guided missile era started about 1947 and extended into the 1960s. The Matador, a ground-to-ground missile, the Snark, a ground-to-ground missile, the BOMARC, a ground-to-air missile, the RASCAL, an air-to-ground missile, and the Falcon, an air-to-air missile, all used a control system of the fly-by-wire type. The experience gained by the manufacturers of electronic components and by users of this equipment afforded an opportunity to establish more technology for fly-by-wire control systems for aircraft. For example, some of the environmental extremes these missile systems had to meet gave the designers the know-how when similar designs were required for manned vehicles with less harsh operational requirements.

In 1956, when *Aviation Week* interviewed personnel of the Air Force Flight Control Laboratory, it was quite apparent that a number of hurdles and roadblocks had to be overcome if fly-by-wire were to become a reality:

> The time may not be far away when the complex mechanical linkage between the pilot's control stick and the airplane's control surface (or boost system valve) is replaced with an electrical servo system. It has long been recognized that this "fly-by-wire" approach offered attractive possibilities for reducing weight

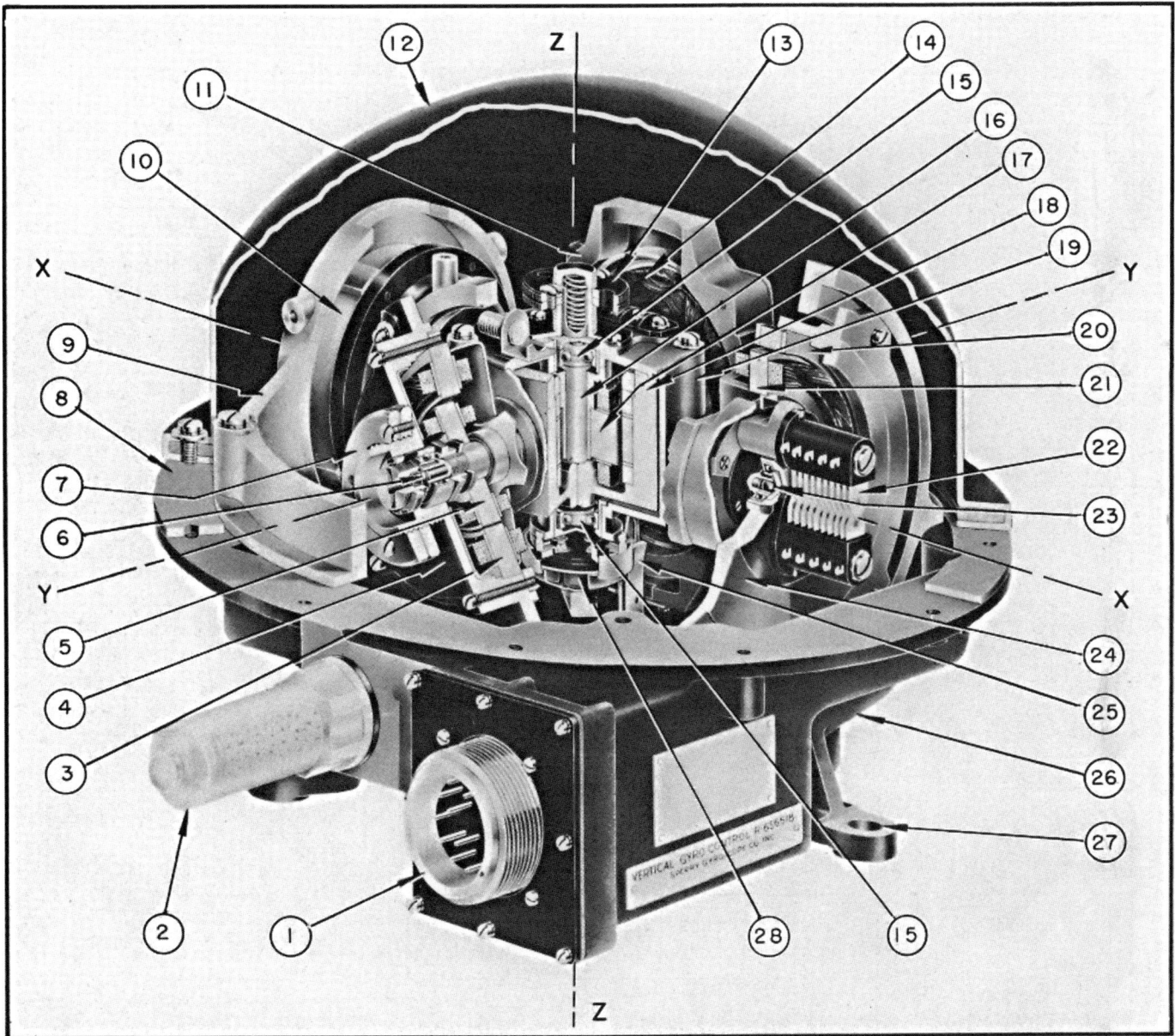

Fig. 1 Vertical gyro control—cutaway view.

1. Electrical Receptacle
2. Dehydrator Plug (Silica-Gel)
3. Pitch-Flight-Selsyn Stator
4. Pitch Flight Selsyn
5. Pitch-Flight-Selsyn Rotor
6. Gyro Ball Pivot
7. Balance Weight
8. Gasket
9. Frame
10. Roll Flight Selsyn
11. Gimbal Ring

12. Cover
13. Balance Weight
14. Roll Erection Torque Motor
15. Gyro Rotor Ball Bearing
16. Gyro Rotor Shaft
17. Gyro Stator
18. Gyro Rotor
19. Gyro Housing
20. Pitch-Erection-Torque-Motor Squirrel Cage
21. Pitch-Erection-Torque-Motor Stator
22. Pivot Contact Assembly

23. Gimbal Ring Ball Pivot
24. Frame End Cap
25. Gyro Tilt Stop
26. Base
27. Mounting Hole
28. Liquid Erection Control Switch

X. Longitudinal (Roll) Axis
Y. Lateral (Pitch) Axis
Z. Vertical (Yaw) Axis

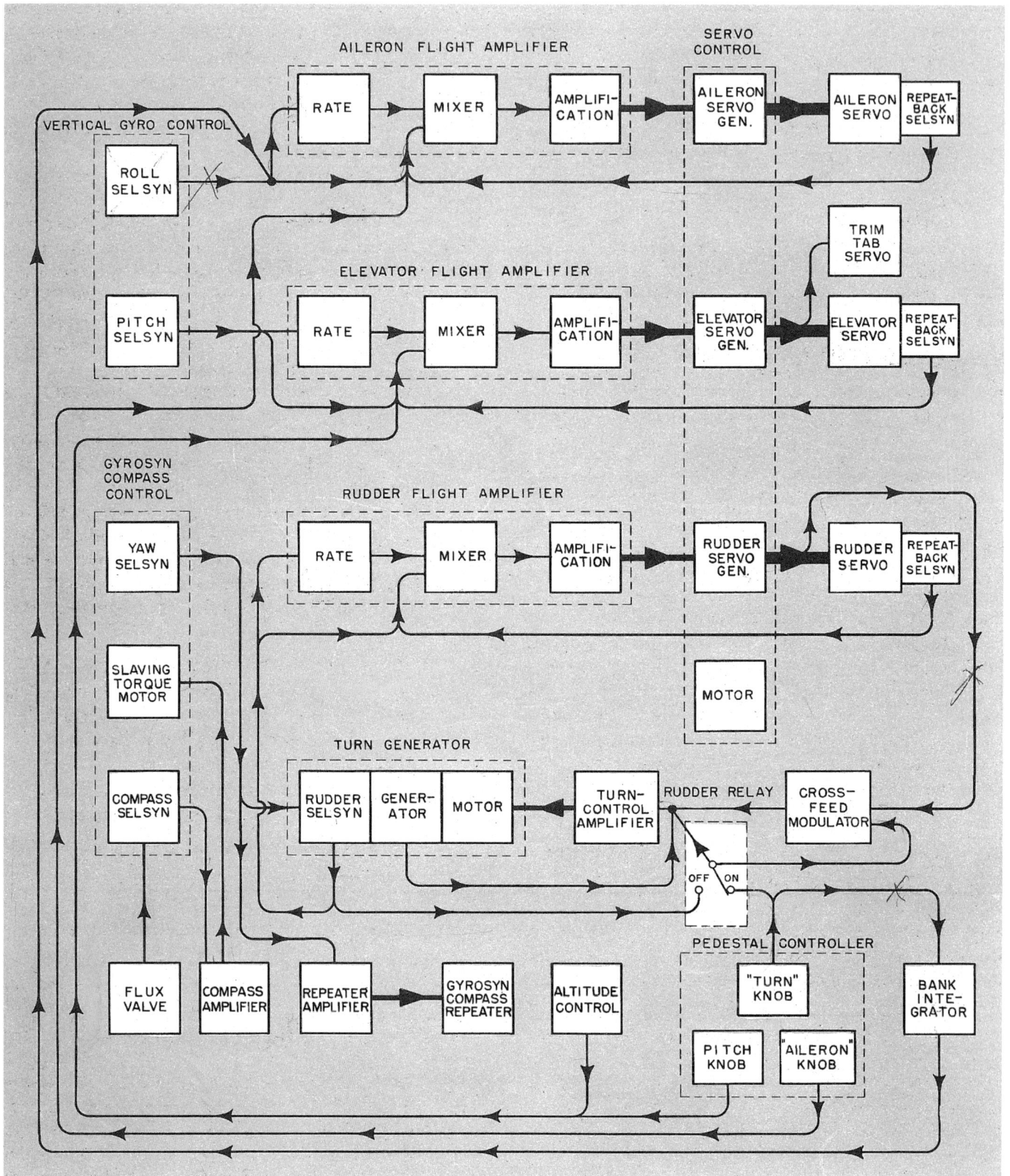

Fig. 2   *A-12 gyropilot—basic signal schematic.*

and complexity. However, airplane designers and pilots have been reluctant to entrust such a vital function to electronics whose reliability record leaves something to be desired.[1]

Thus, the problems were not only technical but also involved individuals' feelings as to what would be acceptable. Providing sufficient data and proof to raise the "confidence level" of all concerned—the designer, the pilot, and the organizations flying and maintaining the aircraft—dictated the type of future programs that had to be undertaken. And, as with all R&D, knowing what must be done is only a part—having the resources and funding was the other part. This aspect of the design program will be discussed later in this book.

Technology in the 1960s

Advantages of Fly-by-Wire over Mechanical Systems

As a design technique for flight control systems, fly-by-wire permits the pilot's command inputs (signals) to be transmitted electrically over considerable distance with fidelity—without distortion, speedily, and with a minimum of energy. The system is easy to install and maintain; it requires no adjustments or devices to ensure proper operation once installed. Signals from multiple sources may be added or subtracted by merely connecting the various conductors together. Although AC signals from various sensing elements could present a phase problem, they all use the same power and hence the same reference voltage.

Electrical systems do not have the problems of friction and wear that occur with mechanical systems. The bending of structural members of the aircraft can affect the mechanical linkages and induce spurious signals to the actuator, but signals transmitted via electrical wires are not affected by bending or vibrations of the aircraft's structural members. Temperature changes have little effect on electrical signaling designs, whereas mechanical systems, especially those using cables and operating over a wide temperature range, are likely to encounter problems.

Flight Control System Designs

The flight control system input/output relationship starts with the input by the pilot and ends by moving an aerodynamic control surface. When advanced and high performance aircraft required more physical effort by the pilot to move the control surfaces, *actuators* (hydraulic types) were employed. New flight control designs, such as the electrically controlled spoilers on the F-111, were necessary because the wing design and its changing position practically mandated the use of electrical over mechanical control of the actuation system.

On the B-47 a hydraulic boost was used: a hydraulic actuator was connected in parallel with the pilot's linkage to move the control surface. Several other aircraft used what was termed a "fully powered" hydraulic control. On this design the pilot's stick, via a mechanical linkage, moved a hydraulic valve on the actuator. The actuator was connected to the control surface and, as the valve was an integral part of the actuator, its output was reflected by movement of the surface. This actuator was called a *moving body actuator* as it moved in accordance with the pilot's input to the control valve.

In a fully powered controller, because the pilot is no longer directly connected to the control surface, the aerodynamic forces are not reflected back to the controls. Such a system is called *irreversible*. With a design that provides the large forces that are required, pilots no longer feel the forces that formerly resisted their motion. The magnitude

---

1. *Aviation Week,* August 6, 1956, pp. 283 f.

of these restrictive forces is in proportion to the speed and altitude of the aircraft. To replace the pilot's lost sensation of resistance, the *feel system* was created. A feel system employs springs and bob-weights to add forces to the controls as they are displaced by stick and rudder.

Added to and made a part of the main actuator was a smaller actuator, typically used to add in "series" and thus termed the *series actuator*. It received inputs from another control valve electrically operated by sensing elements.

When any major changes are made in the flight control system, consideration must be given as to how it may affect the pilot feel system. In a similar manner, when fly-by-wire is used and the mechanical linkage is replaced, the friction, stiction, and inertia are eliminated and again the feel changes, this time in the forward or input path.

As an aid to the pilot, the *stability augmentation system* was devised, consisting of sensing elements, electronic amplifier(s), and a series servo; the output was summed with the pilot's input to the main actuator. Figure 3A is a graphic illustration of this system. The stability augmentation system (SAS) had a limited authority, usually about 5% of pilot's maximum input.

Another system designed to aid the pilot was the *control augmentation system* (CAS). A control augmentation system is a part of the Flight Control System that includes means to shape the command input. Along with sensors and actuators, it performs in such a manner as to augment the static and dynamic stability and maneuvering response of the aircraft. As a separate unit it is essentially a closed-loop tracking control system that responds to pilot commands. A control augmentation system is illustrated graphically in Fig. 3B.

As these systems show, electrical signaling and electronics had been used in some parts of the flight control system before they were used as a primary design technique. With the advancement of aircraft design, mechanical systems were nearing their limits, as the MCAIR Analysis program noted: "The present F-4 system can fairly be described as approaching the practical maximum of reliability achievable in high performance military aircraft with the traditional push rods, torque tubes and cables." (See Fig. 3C.)

A

```
AIRCRAFT
MOTION      →    SERVO
SENSORS          AMPLIFIER
                              LIMITED
                              AUTHORITY
PILOT            MECHANICAL    VALVE
INPUTS           FEEL, FRICTION,  SERIES
                 HYSTERESIS    SERVO    →   VALVE
                                            SURFACE
                                            ACTUATOR
```

B

```
AIRCRAFT
MOTION
SENSORS

                 COMMAND   →  ⊗  →  SERVO
                 MODEL              AMPLIFIER
                                              HIGH
                                              AUTHORITY
PILOT            MECHANICAL         VALVE
INPUT            FEEL, FRICTION     SERIES
                 HYSTERESIS         SERVO   →   VALVE
                                                SURFACE
                                                ACTUATOR
```

C

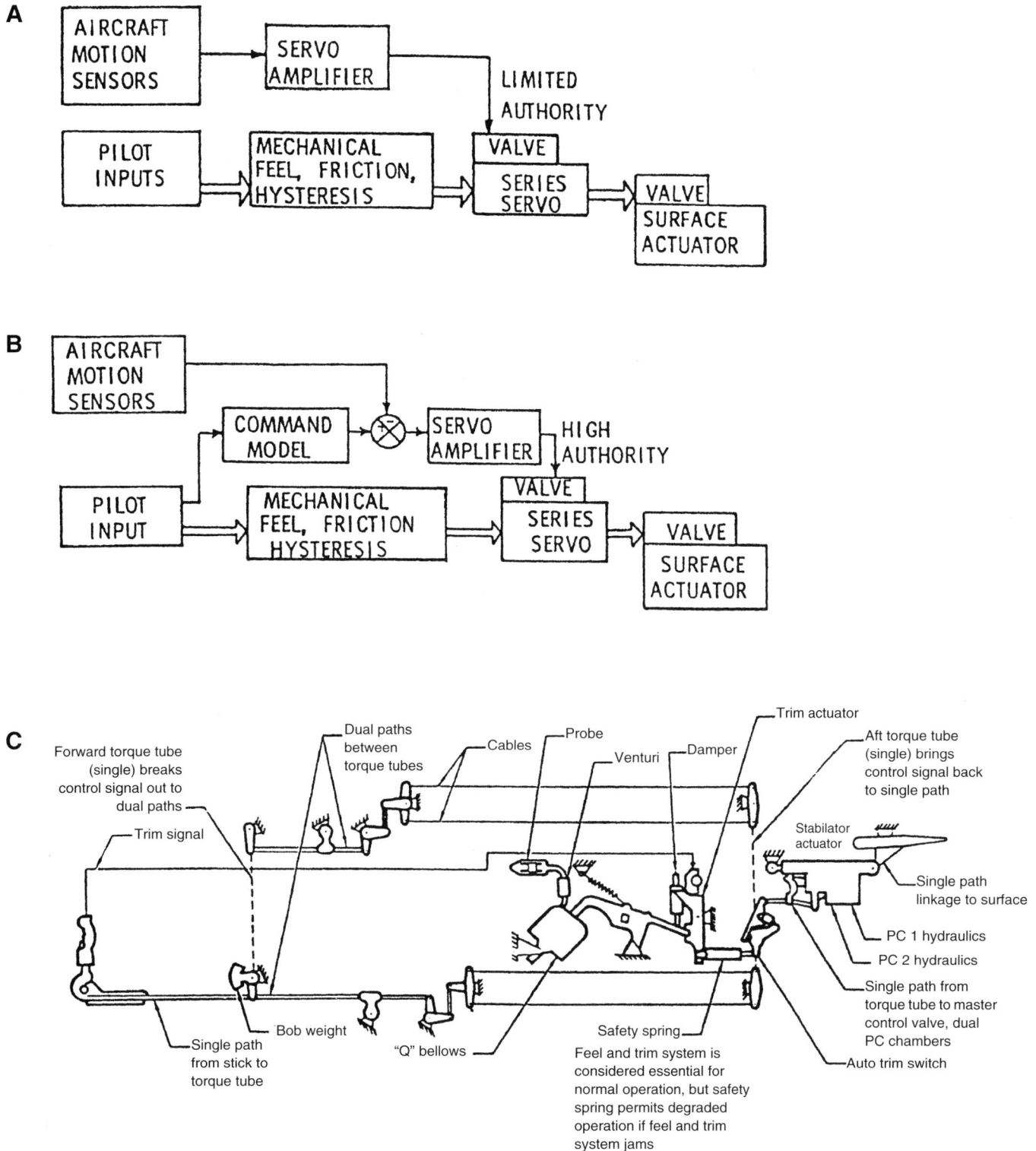

Dual paths between torque tubes — Cables — Probe — Venturi — Damper — Trim actuator

Forward torque tube (single) breaks control signal out to dual paths

Trim signal

Aft torque tube (single) brings control signal back to single path

Stabilator actuator

Single path linkage to surface

PC 1 hydraulics

PC 2 hydraulics

Single path from torque tube to master control valve, dual PC chambers

Auto trim switch

Bob weight

"Q" bellows

Safety spring

Single path from stick to torque tube

Feel and trim system is considered essential for normal operation, but safety spring permits degraded operation if feel and trim system jams

*Fig. 3 (**A**) Stability augmentation (often called "damper"). F4, Century series, C-141.*
*(**B**) Control augmentation (remove mechanical link from pilot to series servo for fly-by-wire). B-70, F-111, A7A.*
*(**C**) Mechanical longitudinal control linkage.*

Chapter 3

Required Programs

Basic Problems and Plans

By 1965 the technical problems with fly-by-wire were fairly well understood. The next step was to translate these problems into work statements and have qualified organizations perform the research to solve them and come up with a viable solution. From the military standpoint, once the funds were available, this meant a Statement of Work had to be prepared. These were sent out for bid. Based on an organization's proposed methods for compliance on the work, a contractor was chosen. Usually, from start of plans to contract took approximately 1 year.

The work by Douglas Aircraft Company at Long Beach, California, had been under way for several years; after several visits by V. R. Schmitt in 1966–68 to review their laboratory model, it was evident that progress was extremely slow. It was obvious from the problems they encountered that other R&D efforts would be required if fly-by-wire were ever to become an acceptable design for flight control systems. The biggest problem was electronics, or rather the application of electronics in fly-by-wire control systems. Therefore, somewhere down the line an R&D effort would have to be undertaken to provide the best solutions where electronics were used.

Other parts and components of the flight control system also deserved attention. For example, one design used during WWII on German aircraft had the actuator packaged with its own power unit. The control and power inputs were electrical; the output arm was connected to the control surface, and thus no hydraulics were required. The concept certainly had potential application, and plans for work on such items were made.

After the Dynasoar (X-20) Project with the Boeing Company was canceled, the laboratory model they used was shipped to Wright-Patterson Air Force Base (WPAFB) and installed in our laboratory. In-house work on this model provided valuable experience for fly-by-wire and actuating systems.

It was also agreed within the Air Force Flight Dynamics Laboratory (AFFDL) that all work on fly-by-wire flight control would be accomplished under one task and one manager. This simplified the planning and directing of the various work efforts that were required. Probably the most significant advantage of the task arrangement was that useful data and test results from one program could be shifted or transferred to another, thus filling voids or overcoming difficulties that otherwise might have taken months of work. For example, the technology derived on the Sperry Phoenix effort was used when the Douglas Long Beach program was redirected.

Although the various R&D programs appear to follow each other in the order they are presented herein, they did not actually occur chronologically. In fact, at one time four projects were under way simultaneously, namely the Douglas program at Long Beach, the LTV program on the Simplex Actuator Package, the Sperry Phoenix program, and the in-house effort on the B-47 aircraft.

Douglas Long Beach Programs

First Douglas Program

The Douglas study began in 1960 with the goal of replacing the mechanical flight control linkage between the control stick and the surface actuators with an electrical link that used no electronics or switching. The specter of unreliable vacuum tubes and early transistors very likely spawned the idea of eliminating electronics. Switching was eliminated also for reliability reasons, just as it is minimized today.

The Douglas program focused on the "hard-wire approach," using one continuous wire from the pilot's station to the actuator over which the control signals would be transmitted electrically. If there were no connections or use of electronics, then the system should be extremely reliable. To ensure a high degree of reliability, three channels were used. Further, as the system would employ no signal amplification, the signal needed to be of sufficient power to drive or control the hydraulic actuator. The transmitted signals were AC, which required a linear variable differential transformer (LVDT) type of transducer at the pilot's station and an actuator design that would accept and convert the AC inputs to control the hydraulic flow of the actuator. In one sense, the type of conversion element dictated the magnitude of the signal to be transmitted.

The state of the art design at that time used two-stage electrohydraulic servovalves mounted on the actuator to control the hydraulic flow to the actuator. These electrohydraulic servovalves operated on DC inputs. In keeping with the Douglas philosophy of using AC only, this required development of AC electrohydraulic servovalves or some design whose performance would be the same or very similar.

The system operated directly from the aircraft's AC power to eliminate any DC conversion equipment. Therefore, the control stick position transducers were LVDTs (linear variable differential transformers) and the hydraulic servovalves used AC torquers. Figure 4 shows a diagram of the system for the pitch axis. The system employed triple redundancy to obtain a reliability equal to the Douglas AD Skyraider pitch control system. Monitoring was performed at the servovalve torquer, which also served as the summing function for the servo input and mechanical feedback.

The Douglas system is shown schematically in Figures 5 and 6. An electromechanical actuator in the actuator's feedback linkage supplies trim. A cockpit display presents the signals from the three servovalve torquers so that the pilot can visually monitor the operation of each axis. The signal from each torquer drives one of three small bars on the display. Under normal conditions, the three bars move together to form a line that moves up and down. When a channel fails, its bar moves away from the other two. The pilot then notes the difference and disables the failed channel by manually operating a switch that places a choke in series with the electrical signal to reduce the signal to a very low value.

This monitoring technique was subsequently judged inadequate because the monitor distracted the pilot's attention from more important flying duties. An automatic failure detection scheme was thereafter devised to eliminate this problem. The scheme compared the torque generated by the servovalve torquer flux against a fixed spring torque.

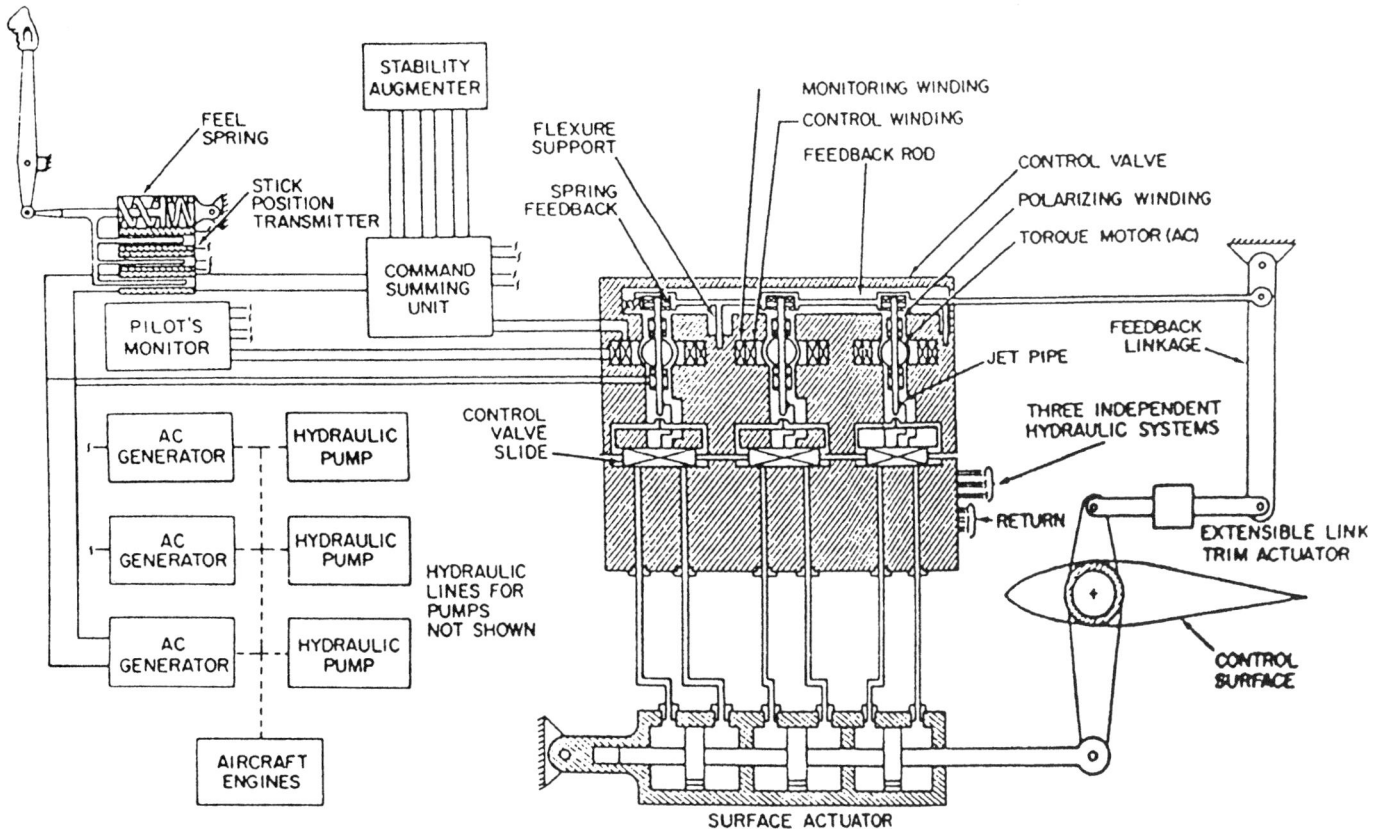

Fig. 4   Schematic diagram—Douglas electrical flight control design.

When a failure caused the flux to exceed 105% of normal maximum, the spring torque was overcome to operate a hydraulic shutoff valve. This scheme was not implemented in the laboratory model, so neither its effectiveness nor the switching time was determined. However, even with the automatic failure detection scheme, failures in the servo-valve second stages would escape detection. The actuator employed three tandem rams and three servovalves having coupled second-stage spools. Active redundancy was employed with all channels connected to the output.

The probability of a total failure of the fly-by-wire system was estimated at 3.15×10^{-4} for a 1.5-hour mission, compared to 6.15×10^{-4} for the original mechanical system. However, the probability of a single failure occurring was 101.7×10^{-4} and 20.1×10^{-4}, respectively. In other words, the fly-by-wire system would incur a total system failure only half as often, but it would require maintenance actions five times as often as the mechanical system.

Although the Douglas study showed that a fly-by-wire system might be designed without either electronics or switching that could match the reliability of a mechanical system, the study and the design had a number of failings:

1. The study failed to include any discussion of artificial feel implementation, which is vitally important to a practical fly-by-wire system.
2. The AC servovalve torquers were very inefficient devices that required a great deal of electrical power from the stick position LVDT for operation, particularly as additional torque was required to operate with the

13

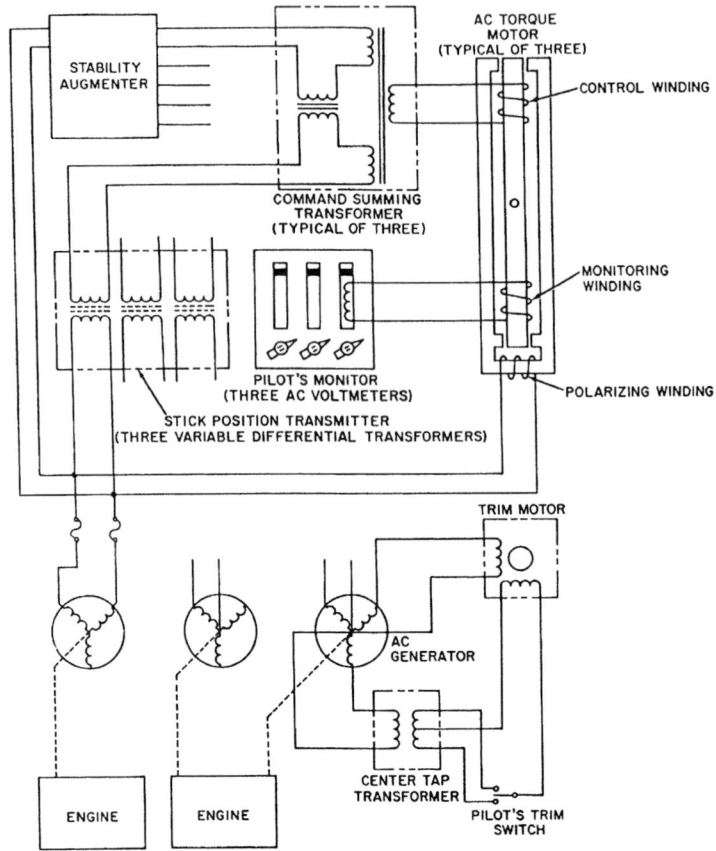

Fig. 5   Electrical schematic—Douglas electrical flight control system.

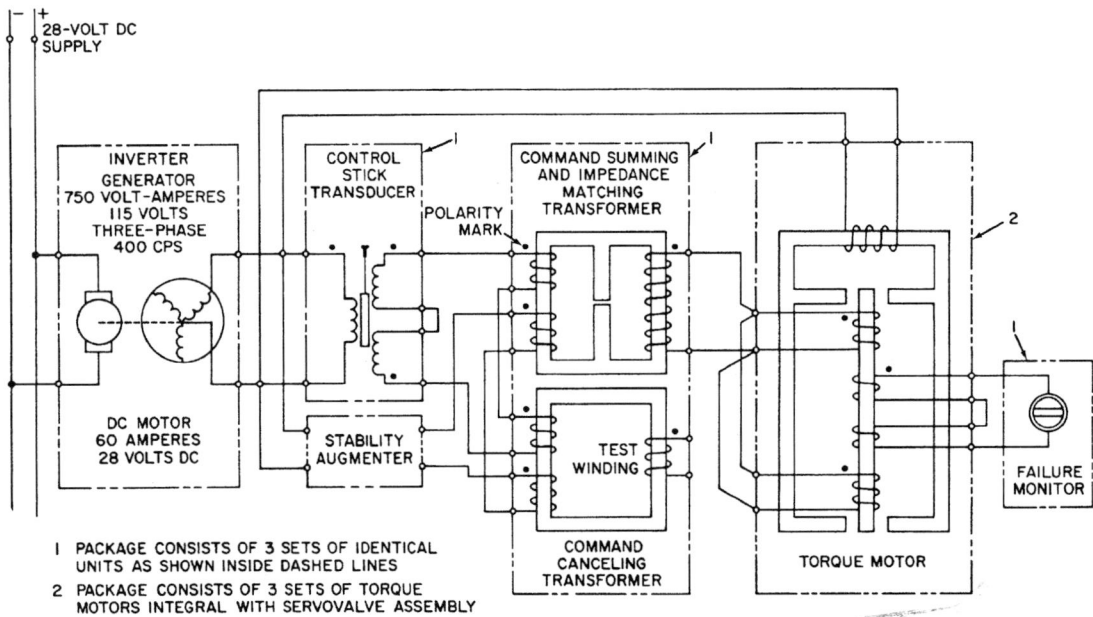

Fig. 6   Diagram of one channel—Douglas electrical primary flight control system electrical circuit.

mechanical feedback. The three valves required a total power of 50 watts. The triplex LVDT absorbed another 60 watts at its maximum displacement.

3. The components were extremely large and heavy, thus partially negating one of the basic advantages of fly-by-wire—size and weight reduction. The breadboard model of LVDT and servovalve (excluding the actuator) weighed 30 and 55 pounds, respectively. Although the flightworthy components would certainly weigh much less than the models, the trend was obvious. (For comparison, a triplex signal LVDT used in electronic flight control designs would weigh about 5 ounces.)

4. The magnetic summing and monitoring techniques were not practical for two reasons. First, signals from different power supplies could not be summed inductively unless they were exactly synchronized; otherwise, the output signal would bear no significant relationship to the desired signal. Second, because the transfer impedance of a transformer depends on the flux level in the core, the output level for one input signal depended on the presence and level of a second input. This nonlinear effect caused a varying forward path gain in the control system.

5. The gradient of surface deflection per control stick displacement was reduced by one-third for each electrical channel failure. One channel failure reduced the command torque at the servovalve input to two-thirds normal, which was balanced by the feedback torque produced by two-thirds normal surface deflection.

The use of mechanical feedback and coupled servovalves presented very difficult design and synchronization problems. At least 2 years were spent in developing a prototype model with only limited success. It was concluded from the program evaluation that the Douglas approach was not suitable for use in fly-by-wire systems. Although the results were negative, the program provided a beneficial contribution to fly-by-wire development because it prevented others from attempting the same approach.

Douglas Redirected Program

After Douglas had shown that the original design concepts would not provide useful hardware the program was redirected. The design requirements on the redirected effort were changed to permit the use of electronics, DC servovalves, and voting schemes on redundant systems. Douglas personnel were asked to review with the Sperry Company of Phoenix, Arizona, their recently developed methods on voting techniques as applied to redundant systems. These techniques had not been available to Douglas during their original program; after discussions with Sperry, Douglas applied one method on their triplex actuator.

The objectives of the redirected Douglas program were as follows:

1. Design, fabricate, and test a laboratory model of a single-axis three-channel fly-by-wire system.
2. Evaluate the performance of this model under standard conditions and simulated failure modes.

Figure 7 is a schematic of the three-channel system. The model was designed such that the electrical and hydraulic power supplies of each channel were independent of the others. Jet-pipe–type electrohydraulic servovalves were used.

A functional block diagram of the system is shown in Figure 8. The modulating piston position is the output of the inner loop; this is also fed to the voter mechanism, which in turn drives the slide valve.

15

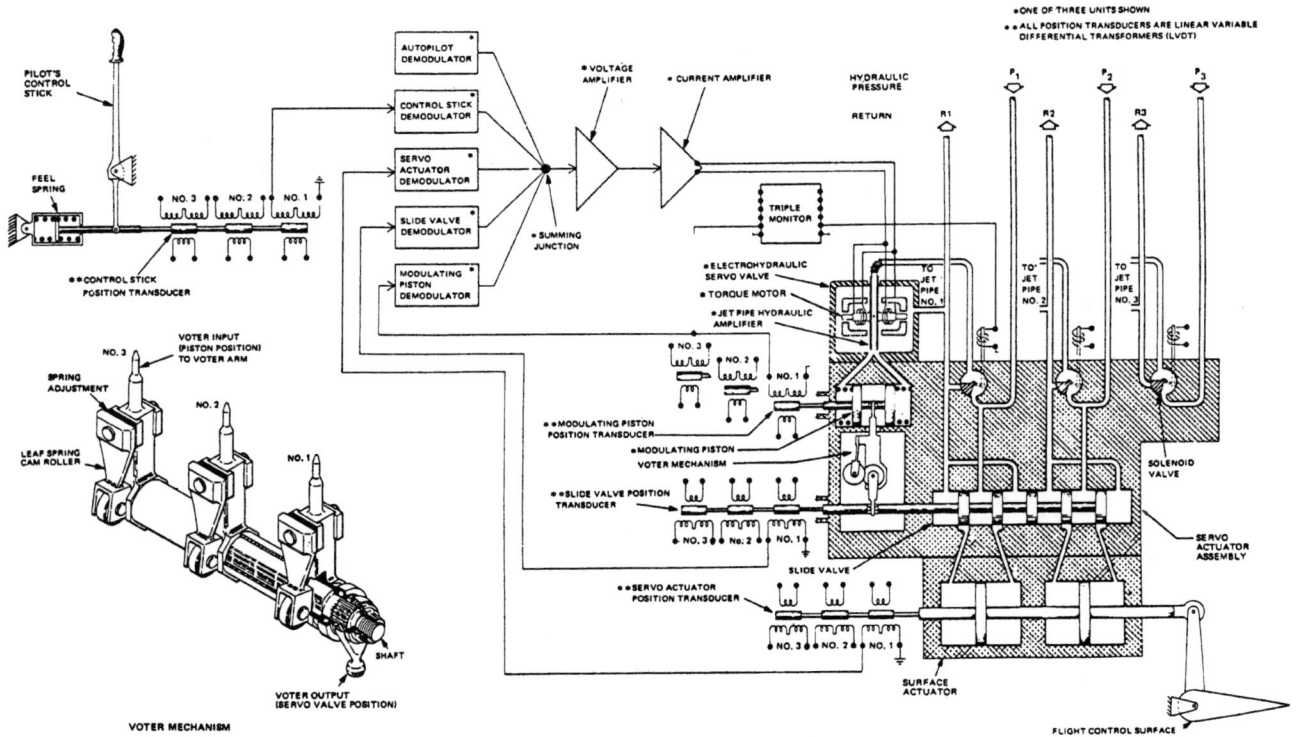

Fig. 7   Fly-by-wire schematic diagram.

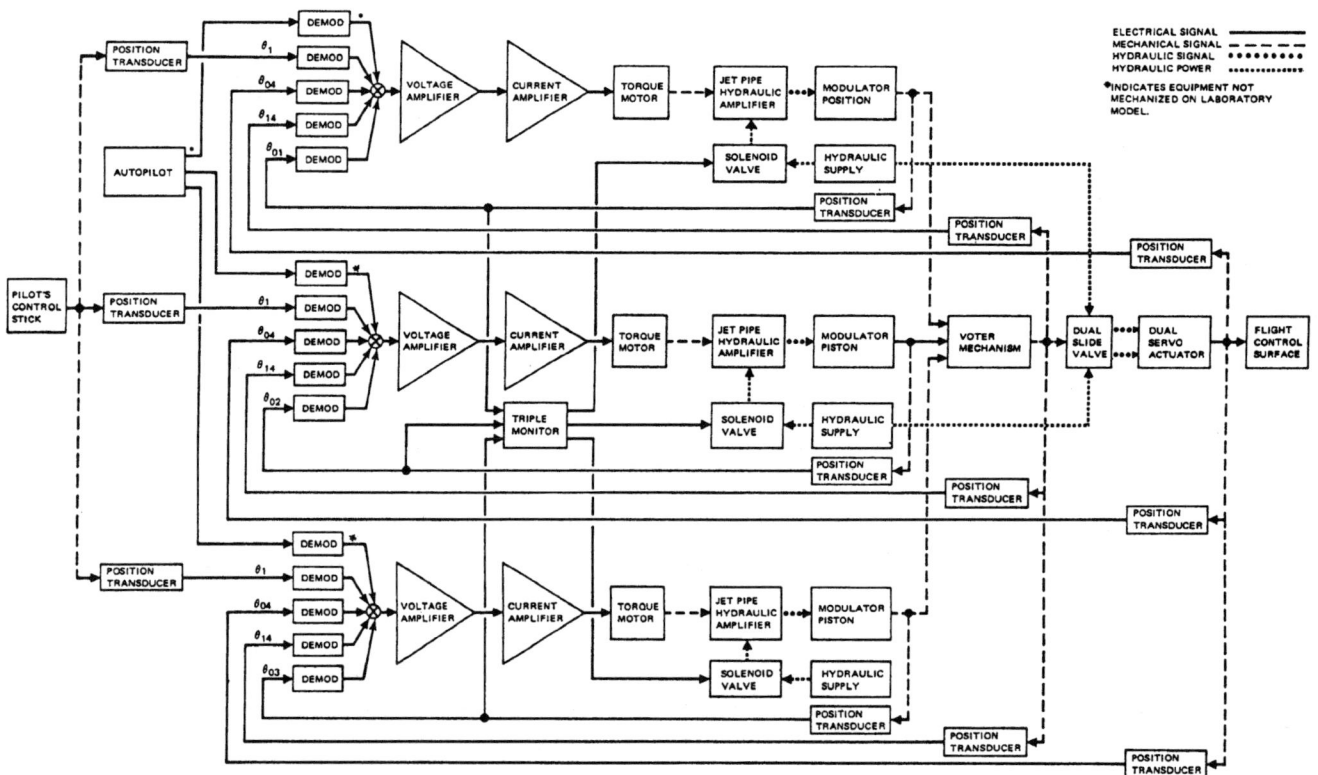

Fig. 8   Fly-by-wire block diagram.

Voter Mechanism

Because the voter mechanism has a major role in the operation of the triplex actuator, a more detailed description seems in order. As shown in Figure 9, it is a mechanical signal selector that continuously selects and transmits a single position from three different mechanical position inputs. Inputs to the voter are the angular positions of the three voter arms; the output is the angular position of the torque summing shaft. The output of this shaft is transmitted to the hydraulic slide valve that ports fluid to the actuator, which in turn drives the aerodynamic control surface.

Two assumptions are made for operation of the voter:

1. The torque on the torque summing shaft due to bearing friction is negligible.
2. The torque on the shaft due to the loading by the slide valve is small compared to the detent breakout force.

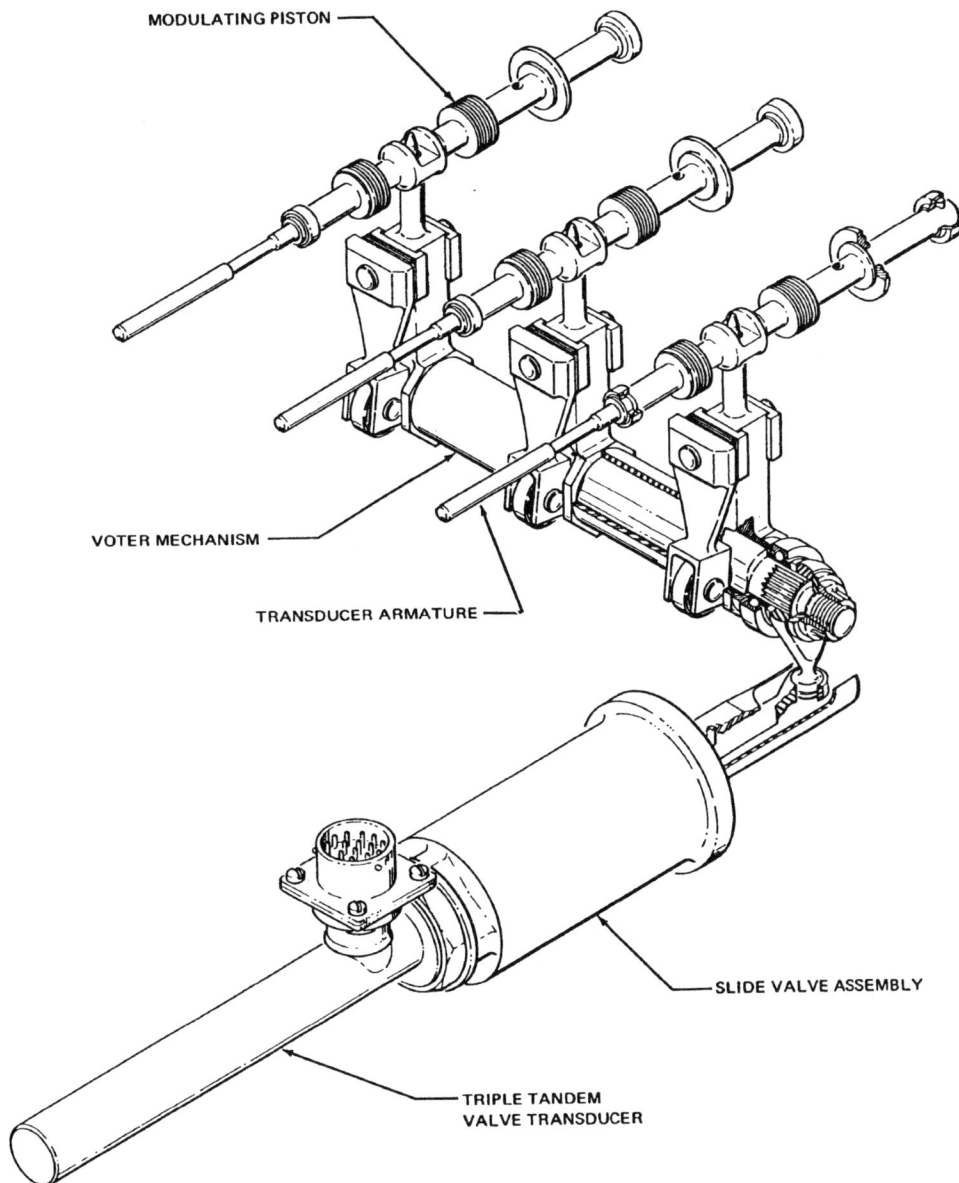

Fig. 9   Voter mechanism arrangement.

17

As the proper values of torque are applied, the voter shaft will be in the position of the midposition voter arm, independent of the positions of the other two voter arms.

The signal from the actuator transducer is fed back to null the input from the pilot's control stick transducer, shown in the photos in Figure 10.

A number of evaluation tests were conducted to determine the system performance, including transients due to channel failures.

Fig. 10 (A) Servoactuator assembly. (B) Control stick transducer assembly.

Test results on this laboratory model demonstrated that a single pilot's command signal can be converted into three signals, transmitted electrically over three independent channels (each of which has a mechanical output), and reconverted by a method of continuous voting into a single mechanical output. This continuous voting principle offered a potential for improving the reliability on fly-by-wire systems.

Performance of the laboratory model of the fly-by-wire system after occurrence of a single electrical failure surpassed the objective set for a fail-operative system. Adequate performance was demonstrated on introduction of a hardover signal, and on complete loss of electric power, hydraulic power, or actuator position feedback in any one of the three channels.

Wright-Patterson Air Force Base B-47 In-House Program

The Wright-Patterson Air Force Base (WPAFB) program covers a time period from August 1966 to November 1969. "In-house" as used here means all work was performed at WPAFB using Air Force facilities, aircraft, and pilots. However, the design, installation, and flight tests of fly-by-wire equipment were accomplished by contractor personnel.

A number of factors led to the choice of a B-47 aircraft for the in-house work on fly-by-wire. The B-47 afforded plenty of room to make installation of equipment and operation of equipment during flight tests. Because it was a multi-seat aircraft, the pilot did not have to divide his time between flying the aircraft and changing dials on test equipment; tests could be accomplished by another crew member, thus maintaining flight safety. The B-47 was a larger aircraft, so the advantages of fly-by-wire over long mechanical linkages or cables might be more easily demonstrated. Also, the B-47 was available as an on-base aircraft for flight testing, so crew and maintenance personnel were always on hand.

Figure 11 is a photo showing the B-47 aircraft used on the program along with contractor and Air Force personnel: (left to right) D. Bazill, G. Jenney, W. Talley, H. Schreadley, C. Black, V. Schmitt, Lt. J. Ramage.

The in-house program was conducted in three phases. Phase I used an electrical nonredundant primary flight control system with control inputs being generated by the normal control column motion. Phase II added a side-stick controller, pitch rate and nose acceleration feedback, and electrical roll axis control to the Phase I system. Phase III incorporated a four-channel redundant actuator with hydraulic logic into the Phase II system.

Phase I

During Phase I, the electrical control channel was installed in parallel with the standard hydromechanical elevator control system. A position transducer attached to the bottom of control column linkage generated an electrical signal that was fed to a servoamplifier (see Figure 12). The servoamplifier output controlled an electohydraulic servovalve that directed flow to the hydraulic actuator attached to the elevator control surface. Thus, as the pilot moved the control column, the elevator moved accordingly. Figures 13 and 14 show the installation of the fly-by-wire components and the cable system to the actuator.

When the fly-by-wire actuator was used, the mechanical actuator was bypassed so that its force output was zero, even though the control cables continued to operate its control valve. Because of the nonredundancy of the electrical channel, the mechanical system was used as a backup.

Fig. 11   B-47 aircraft with contractor and Air Force personnel.

Test Results

Four different test pilots accumulated more than 40 hours of flight time, including touch-and-go landings, on the fly-by-wire system during the Phase I program. Major Fredericks, USAF, Test Director and Test Pilot, evaluated the system as follows:

Handling Qualities   Noticeable improvement in handling qualities offered by the test system was a decrease in backlash due to cable stretch. The fact that the test system did eliminate a measurable amount of cable stretch was demonstrated by trimming the aircraft up with the test system engaged, changing the airspeed without re-trimming, and observing the trim change caused by reverting to the normal aircraft elevator system. Elimination of backlash due to cable stretch made for much more positive, less sloppy, and apparently more responsive elevator control. This was most apparent in high speed, low altitude flight.

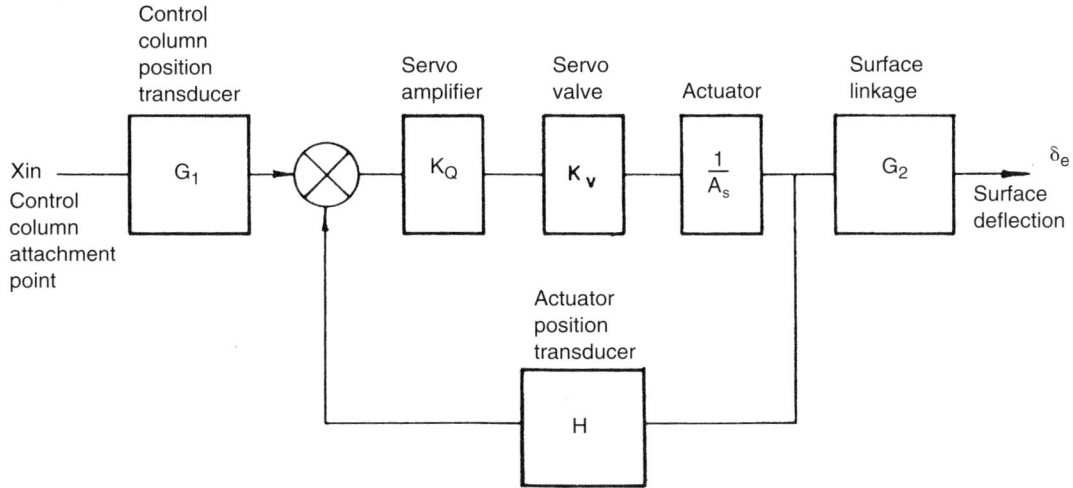

Where G_1 = 0.820 volts/in
       K_a = 84.4 ma/volt
       K_v = 0.623 cis/ma
       H = 0.820 volts/in
       A = 2.7 in^2
       G_2 = 12°/in

Fig. 12   B-47 control circuit block diagram.

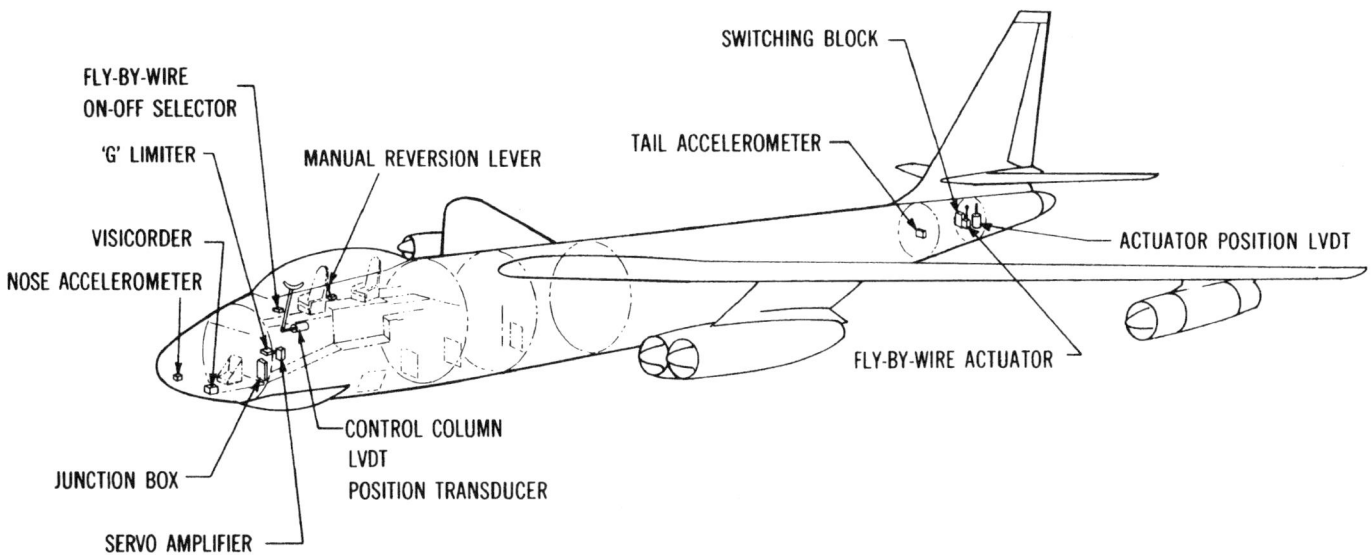

Fig. 13   Phase I system components.

TYPICAL TENSIONER ASSEMBLY

⑤ REF.

TURNBUCLE

⑥

SWITCHING BLOCK

1. DART SENSOR POWER SUPPLY
2. MANUAL REVERSION LEVER
3. FRONT CABLE TENSIONER
4. REVERSION CABLE
5. REAR CABLE TENSIONER
6. FORCE RELIEF UNIT
7. DART RATE SENSOR

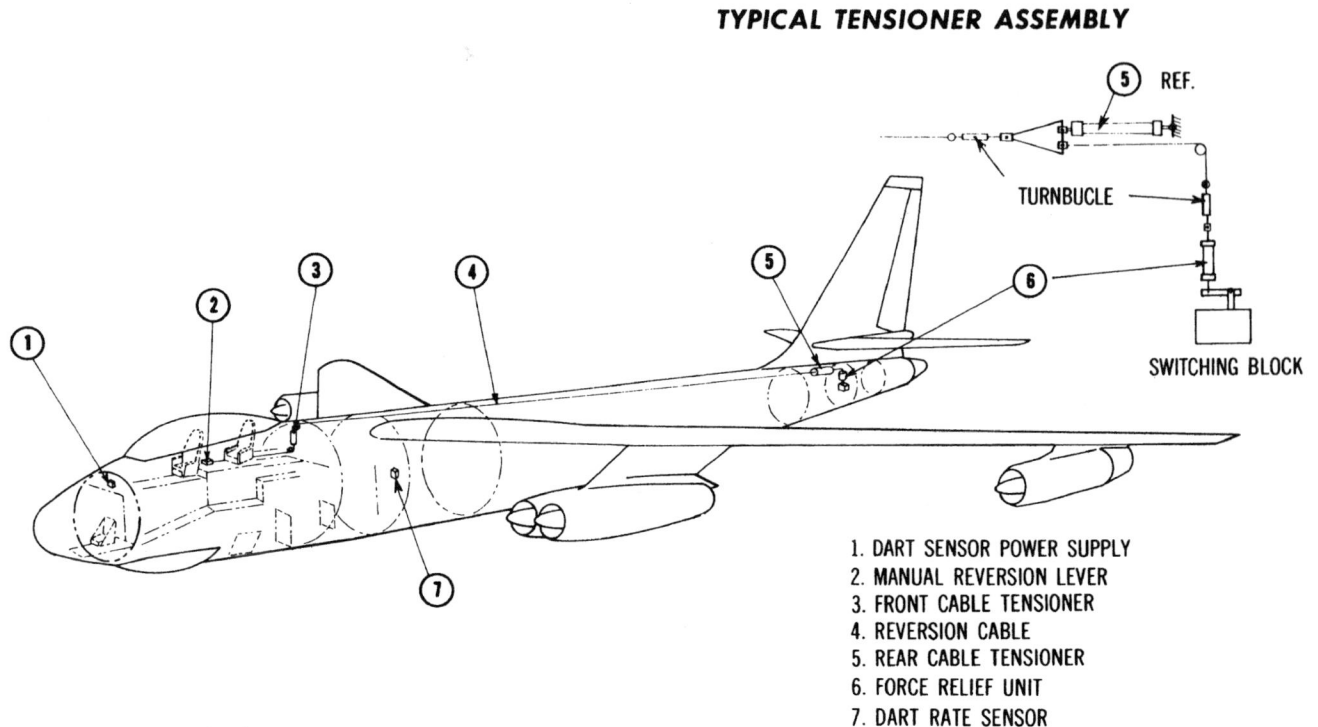

MODIFICATIONS:
MOD 1 CABLE TENSIONING SYSTEM
MOD 11 DART RATE SENSOR

Fig. 14   Rate gyro and cable tensioning system.

System Reliability   The system tested consisted of a simple open loop electrical connection of the control column and the elevator actuator by means of linear transducers and electrical wiring. This system operated in parallel with the normal aircraft elevator control system; both mechanical and electrical means were provided for switching between the normal and test systems. No malfunction in the basic *Fly-by-Wire* system occurred during the entire period of testing. A high degree of confidence in electrical flight control systems prevails, based on experience with this program.

Phase II

The Phase II system used a side-stick controller and pitch rate and nose acceleration feedback in combination with the components of the Phase I system. Roll control by the side-stick was accomplished by driving the modified aircraft autopilot channel and adding rate feedback.

Side-stick Controller

Pilot inputs to the fly-by-wire system were made on Phase II by use of a side-stick controller. The motion transducer and grip were mounted on a slide-tray platform attached to the left armrest of the pilot's ejection seat. A grip was

Fig. 15   Side-stick assembly incorporating elbow cup and roll trim knob.

fabricated that incorporated a pitch trim knob and a fly-by-wire engage and disengage push-button switch (see Figure 15). The pilot also controlled roll by use of the side-stick (see Figure 16). The side-stick was adjusted for displacement as follows:

Displacement: Pitch Axis, ± 13; Roll Axis, $\pm 17°$

C* Description and Criteria

"C star," (written as a capitalized letter "C" with an asterisk) is an expression related to the handling qualities of an aircraft. Mechanization of the basic pitch axis of Phase II was a side-stick controlled, acceleration and pitch rate feedback blend system, as shown in Figure 17. The C* criteria, generated by Tobie, Elliot, and Malcom of the Boeing Company in 1965, is based on the idea that the "pilot responds to a blend of pitch rate and normal acceleration, with the blend ratio varying according to the normal variations in aircraft response."

23

Input Filter — $\dfrac{K_8}{T_8 S + 1}$

Control Surface and Servoactuator* — $\dfrac{K_9}{T_9 S + 1}$

Roll Rate — A/C DYNAMICS $\dfrac{\dot{\theta}}{\delta a}$ Transfer Function

Pilot's Input (Roll) — Volts — V+ — V — δa Degrees — $\dot{\theta}$

Filter — $\dfrac{K_{11}}{T_{11} S + 1}$ — Volts

Rate Gyro — K_{10} — Volts — rad/sec

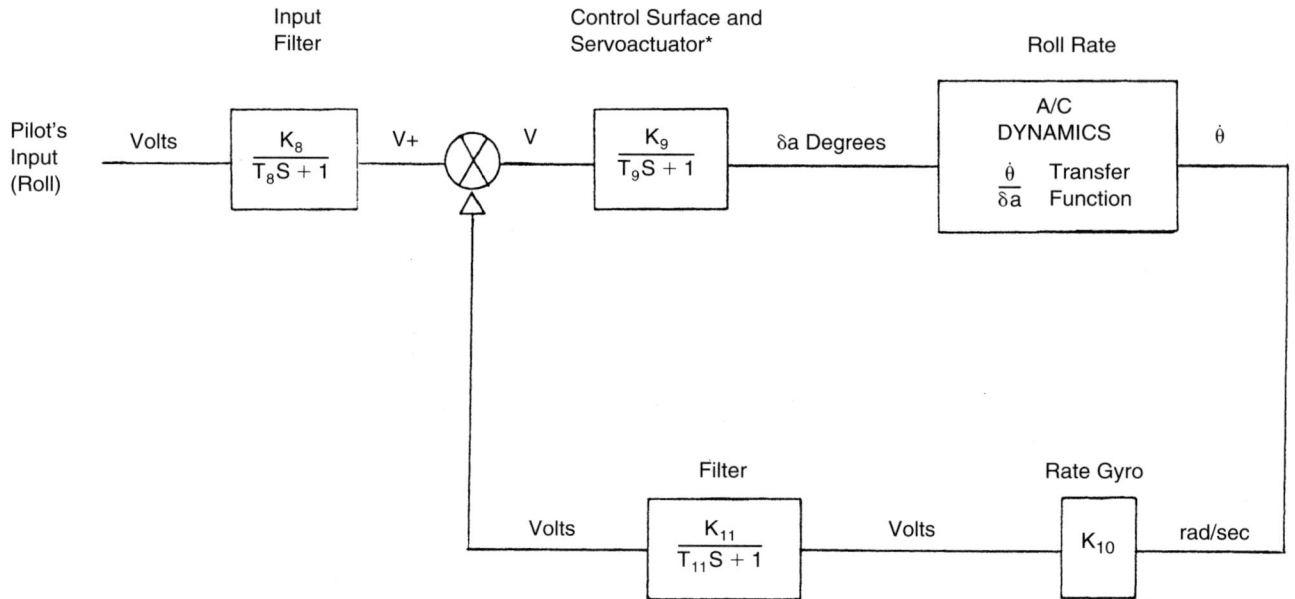

* Approximation for small angles. Controlled by autopilot servo system.

Fig. 16   B-47 Phase II control system roll axis block diagram.

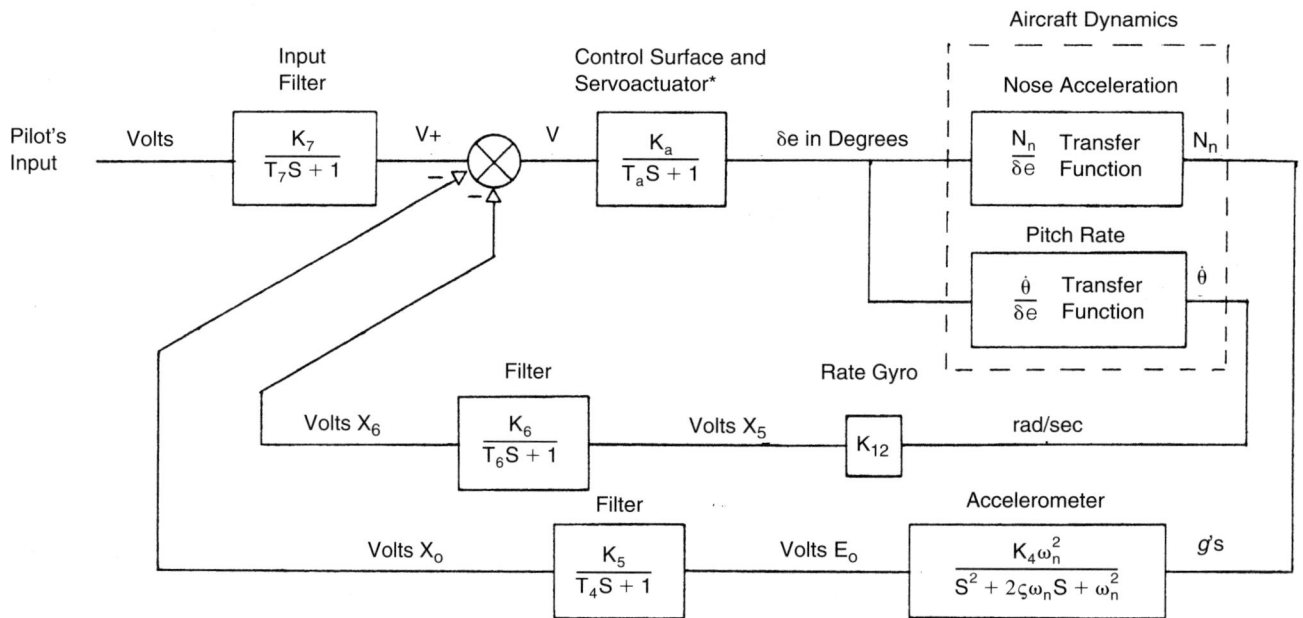

Aircraft Dynamics

Input Filter — $\dfrac{K_7}{T_7 S + 1}$

Control Surface and Servoactuator* — $\dfrac{K_a}{T_a S + 1}$

Nose Acceleration — $\dfrac{N_n}{\delta e}$ Transfer Function — N_n

Pitch Rate — $\dfrac{\dot{\theta}}{\delta e}$ Transfer Function — $\dot{\theta}$

Pilot's Input — Volts — V+ — V — δe in Degrees

Filter — $\dfrac{K_6}{T_6 S + 1}$ — Volts X_6 — Volts X_5

Rate Gyro — K_{12} — rad/sec

Filter — $\dfrac{K_5}{T_4 S + 1}$ — Volts X_0 — Volts E_0

Accelerometer — $\dfrac{K_4 \omega_n^2}{S^2 + 2\varsigma\omega_n S + \omega_n^2}$ — g's

* Approximation for small angles.

Fig. 17   B-47 Phase II C control system pitch axis block diagram (I).*

The relationship between N_Z and V may be derived using Figure 18. As the aircraft pitches up at angle θ, it attempts to pivot around a point P. The rate of angular rotation around P is $\dot{\theta}$, where

$$\dot{\theta} = \frac{d\theta}{dt} = \omega$$

However,

$$V = R\dot{\theta}$$

where

  R is the radius of the pivot point from P.
  V is the airspeed of the aircraft.

The normal centrifugal acceleration N_Z is equal to

$$R\omega^2 = R\dot{\theta}^2$$

Combining these relationships yields

$$\frac{N_Z}{\dot{\theta}} = V$$

Thus, for a "perfect" pitch response, the ratio of acceleration felt by the pilot to the visible rate of change of the position of the horizon varies linearly with airspeed.

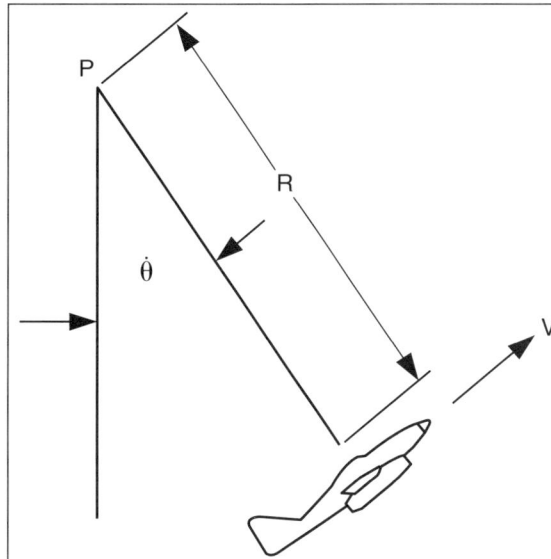

Fig. 18   Diagram for derivation.

C Equations*

$$C* = K_1 N_Z + K_2 \dot{\theta} \qquad \text{(pilot at c.g.)}$$

$$C* = K_1 N_Z + K_3 \ddot{\theta} + K_2 \dot{\theta} \qquad \text{(pilot not at c.g.)}$$

where

c.g. is the aircraft center of gravity.

K_3 is the distance from the pilot to the center of gravity divided by 1 *g*.

g is acceleration of gravity (32.2).

C Design Procedures*

The procedures for establishing the gain blend for the θ and N_Z feedback requires selecting a crossover velocity, V_{CO} (the velocity at which $\dot{\theta}$ and N_Z cues have the same magnitude).

At crossover: $K_1 N_Z = K_2 \dot{\theta}$

where

$$\dot{\theta} = \frac{N_Z(g)\ (32.2)}{V_{CO}\ \text{(feet/second)}}$$

let

$$K_1 = 1$$

then

$$K_2 = \frac{V_{CO}}{32.2}$$

For the B-47, a $V_{CO} = 400$ feet/second was used, thus giving

$$\frac{K_2}{K_1 1} \text{ ratio of } 12.4$$

This implies that the steady-state ratio of volts/radian/seconds to volts/gravity should be 12.4 for a crossover velocity of 400 feet/second.

• **Pilot Pitch Input Shaping**   To smooth the pilot's pitch inputs, a first-order lag filter having a selectable time constant of 0.12, 0.24 and 0.48 seconds was used (see Figure 19).

• **Rate Gyro and Accelerometer Filter**   The accelerometer and rate gyro feedback gains were adjusted for various values during flight. A lag filter with a time constant of 0.0056 seconds was used to eliminate noise from the rate gyro. A lag filter with a time constant of 0.25 seconds for the accelerometer gave the best stability for the desired C* response.

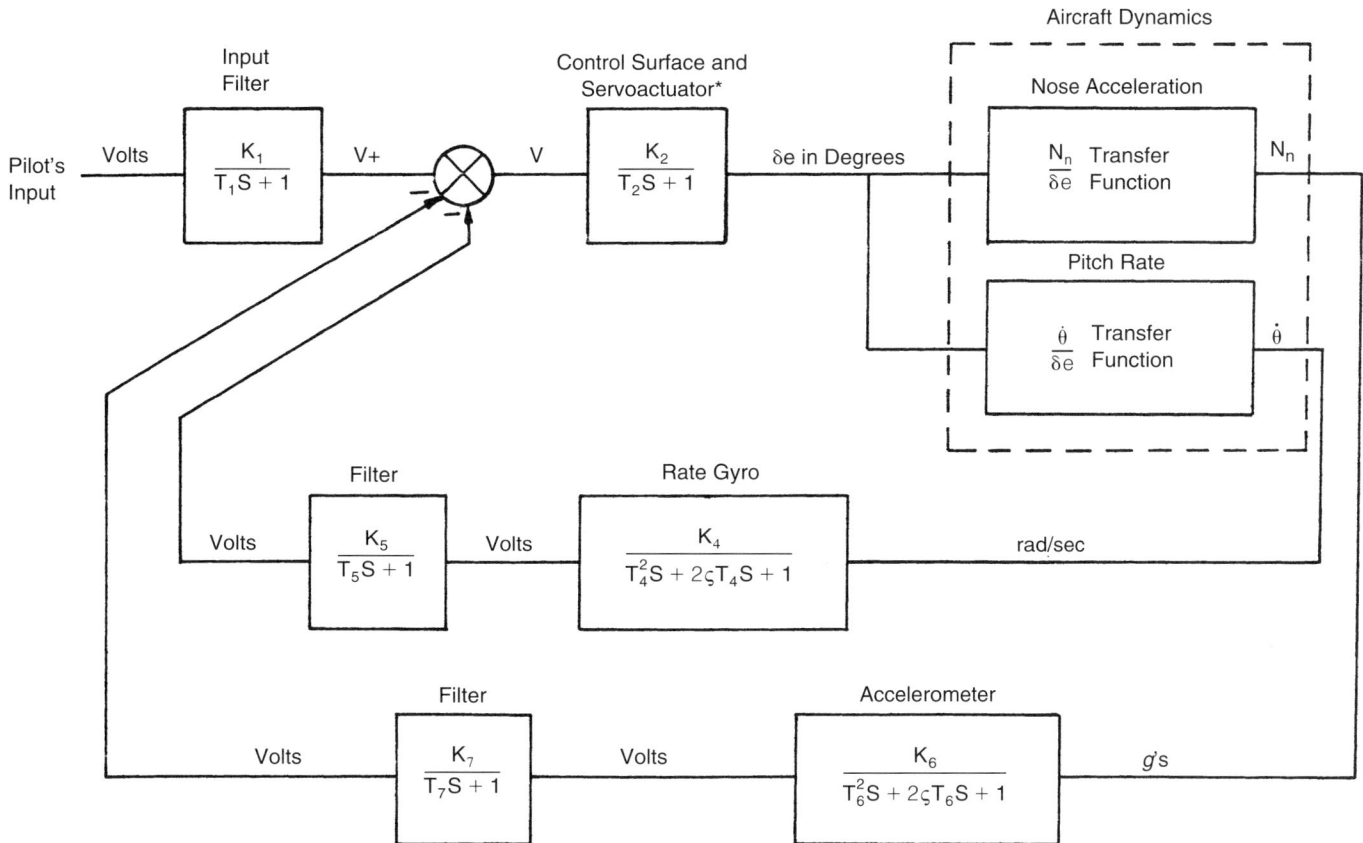

Fig. 19   B-47 Phase II C* control system pitch axis block diagram (II).

Test Results

Col. Giesler, the B-47 test pilot on Phase II (over 2500 hours as B-47 pilot), made a number of comments on the Phase II fly-by-wire flight tests:

> In ease of control there is absolutely no comparison between the standard system and the fly-by-wire. The fly-by-wire is superior in every aspect concerning ease of control. At medium *g* the fly-by-wire is a much better system, a more precise system to fly. . . . You could point your finger and hit just about what you were pointing at; the B-47 on fly-by-wire is like pointing your fingers to fly the airplane. We made landing approaches on a turbulent day with fly-by-wire and there was no effect involved. It is positive, it is rapid—it responds well—and best of all the feel is good. I have never flown an airplane that had such ease of corrections on final approach—and especially the B-47.

Phase III

Objective of Phase III was to evaluate a quadruple-redundant (two fail-operate) electrohydraulic actuator in the B-47 aircraft. The test actuator replaced the nonredundant pitch axis actuator used in the Phase II mechanization. The actuator design was based on the use of self-contained hydraulic logic for the functions of monitoring and failure removal. The unit had active standby redundancy, and the control channels were selectively coupled to the output one at a time. Control channels were similar, so there was no performance change with failures. The actuator could withstand two similar or dissimilar failures (hydraulic supply, electrically–internally or input) with no performance change. If a third channel failed, the output went to a rate-damped soft condition allowing the control surface to move gradually to a trail position.

The actuator was powered by four independent electrically driven hydraulic power supplies. See Figure 20 for the actuator schematic.

Test Results

Failures were injected into the actuator, and the failure removal characteristics were recorded during flight tests. In level flight the failure efforts were practically immeasurable. The effect of failures during aircraft maneuvers was within acceptable limits.

Test pilots stated that the modified side-stick controller for Phase III operated satisfactorily. The normal aircraft elevator system was carried along as a backup for the test system. Handling qualities on fly-by-wire in Phase III showed an improvement over the standard aircraft.

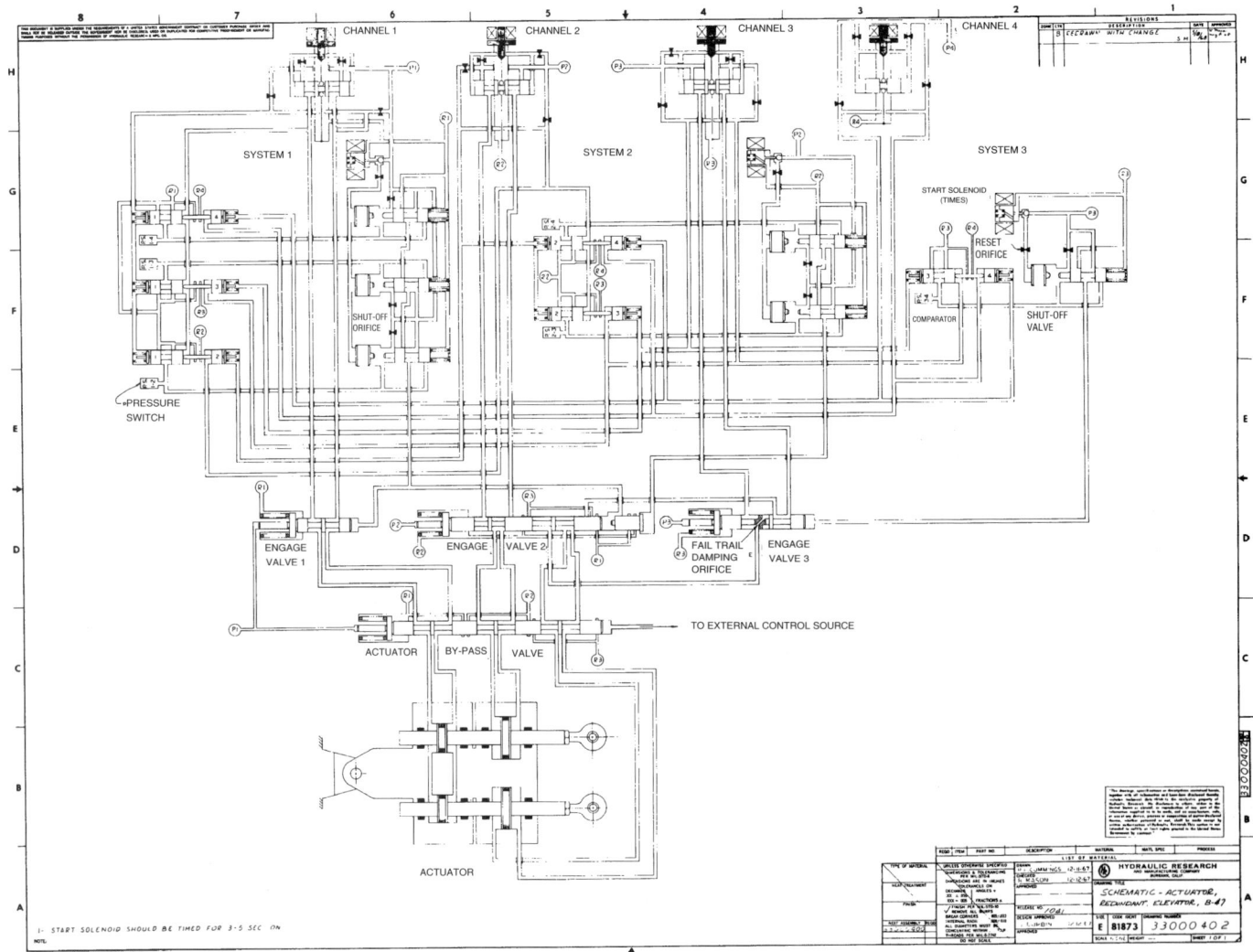

Fig. 20   Quadruple-redundant fly-by-wire actuator schematic.

Fig. 21   C* computer.

WPAFB B-47 Program Results

The B-47 program, by providing positive test results, was a milestone in establishing that fly-by-wire was a viable technique for use in the design of aircraft flight control systems.

This in-house program was first in demonstrating the basic aspects of fly-by-wire on an aircraft:

1. Transmission of electrical control signals.
2. Attainment of handling qualities improvement.
3. Use of redundancy methods for flight safety.

These improvements were accomplished by the design, installation, and flight test of the mechanized hardware:

1. Control by the pilot column and side-stick controller to actuator.
2. C* computer (Figure 21) and two-axis control by pilot.
3. Quadruple-redundant hydraulic actuator.

The positive test results were provided by recorded flight data, and statements by test pilots affirmed that fly-by-wire handling qualities were superior to those of standard aircraft. In addition to increasing the confidence level for using fly-by-wire, the B-47 program provided a fundamental building block for the expanded Advanced Development Program.

LTV Aerospace Integrated Actuator Package Program

Objective

The initial objective of the LTV program was to establish design techniques and determine the feasibility and advantages of using integrated power control servoactuator packages in the flight control system of present day military fighter-bomber aircraft.

After 1969, when data was obtained from combat experience during the Vietnam conflict, the program's objective remained the same but some of the work items were changed. Combat data indicated shortcomings in the aircraft's ability to survive intense ground fire, including fire from small arms. A major contributor to this vulnerability was the flight control system. Thus, the work effort was expanded to determine the degree of survivability of integrated actuator packages (IAPs) over the conventional flight control design.

Pre-1969 Background

The use of integrated packages, either mechanically or electrically signaled, was not completely new. During World War II the Germans employed them in their V-2 rocket and in several of their military aircraft. Electrically powered and controlled IAPs were used as parallel surface actuators for autopilot directional control on the HE-11, JV-88, ME-110, and DO-17. The British IAP design (which started approximately 1954) was used on the VC-10. It was equipped with integrated actuators on each of the split surfaces of all three axes. The initial block of Regulus I missiles built in the late 1940s by Chance Vought (later LTV Aerospace) for the U.S. Navy was equipped with electrically signaled integrated actuator packages built by Vickers. These IAPs used variable-pressure, variable-flow pumping units, with a loop closed around the pump displacement element (servo pump). Subsequently, the integrated package concept was successfully applied to thrust vector control actuation of the Polaris missile.

Based on past experience, the problem in designing IAPs was not one of developing a completely new technology or inventing new components, but rather one of determining the feasibility of applying the concept to present and future high performance attack/fighter and fighter-bomber aircraft, and of upgrading component performance.

1. Methods and materials had to be devised to overcome the heat dissipation problem.
2. The problem of space limitation for installation of IAPs in the wings (lateral axis) had to be resolved.
3. Components having higher performance and temperature capability than those currently available had to be designed.

Definitions

The IAP development program either produced or used a number of terms that should be defined.

Integrated Actuator Package (IAP)   IAP refers to a broad class of flight control actuators with self-contained hydraulic power supplies. Each power supply consists of an electric motor driving a hydraulic pump with reservoir, check valves, filter, relief valves, and associated hydraulic circuitry. Power for the IAPs is derived from the aircraft electrical system.

Simplex Integrated Actuator Package   A simplex IAP is essentially a package containing a single nonredundant actuator and a single motor pump unit with associated hydraulic circuitry. The simplex package as described herein contains the components and functions required for the present F-4 stabilator actuator. In addition, it contains a nonredundant motor pump, reservoir, and monitoring and switching functions integral with the actuator as an emergency backup of the hydraulic power supply.

Duplex Integrated Actuator Package   The duplex IAP has two independent electric-motor–driven hydraulic power supplies, each supplying the hydraulic power for one-half of a dual tandem actuator. To meet the two fail-operate, fail-safe requirements for signal transmission, the package is equipped with quadruply redundant FBW input signal channels.

Power-by-Wire (PBW)   Power-by-wire refers to the transmission of power from the aircraft engine to the control surface actuator by electrical means rather than hydraulic. Instead of generating hydraulic power at the engine accessory pad, electrical power is distributed to IAPs located at the control surfaces, where the hydraulic power is generated.

Funk Strut   A funk strut is a bi-directional spring cartridge that is preloaded and acts as a solid link until the load in the link reaches the preload level, at which point the spring collapses at its spring rate.

Overall Program Data

The conflict in Southeast Asia that began in 1969 revealed a serious shortcoming in high-performance military aircraft then in use, namely the vulnerability of the aircraft flight control system to small arms or ground fire. The Integrated Actuator Package (IAP) concept appeared to be a potential solution to this critical problem. IAP designs used electrical power, thus eliminating hydraulic lines for flight control purposes. The electrical power and signal control lines could easily be made redundant. The threat of hydraulic fire was greatly reduced, permitting the aircraft to sustain considerable battle damage and still remain operational.

Early in the program it was determined that an IAP was feasible to design and construct; however, the IAP concept meant more than merely the feasibility of building a unit. It meant determining to what extent it could be applied to modern high performance aircraft. Because the pitch axis appeared to be the most critical, and cost and time were considerations, only the pitch (longitudinal) axis was investigated on this program. Studies of the application of the IAP concept revealed that problems in packaging, heat dissipation, and motor pump efficiency which had not previously bothered the flight control designer were now paramount concerns. To properly size an IAP—one that would provide the proper stiffness, force output, and response, yet use minimum power, have a high efficiency, and be minimum in weight and size—it was necessary that the relationships between actuator and aircraft aerodynamics be thoroughly studied. In other words, establishment of precise and detailed IAP design requirements required that the actuator output over the entire flight regime be given a close look. For example, special attention and study were devoted to actuator area as related to flutter, control surface slew rates as related to maneuvers and landing, electrical power versus hydraulic power, and component reliabilities.

As an expedient measure, data on the A-7D was used in the study of actuator–aircraft functional relationships. The results of this study were not used per se in a particular design, but they did provide background experience and guidelines for an IAP design once a vehicle was selected.

Investigation of the application of the IAP to modern U.S. operational fighter-bomber aircraft was limited to the F-4 and F-111. The initial study was conducted in sufficient depth that a determination could be made as to the size, cost, ease of installation, and maintainability of an IAP for use in the longitudinal axis of these aircraft. Based on the results of this study, the F-4 aircraft was selected, and the experimental laboratory model IAP, which had been planned if the concept proved feasible, was designed to the F-4 requirements.

Some 5 months after program start, the Air Force urgently needed flight test data applicable to advanced aircraft such as the F-X (now the F-15). Therefore, it was decided to build a flightworthy simplex package instead of a laboratory model. Although this did not change the basic objectives of the program, it added considerably to the amount of design detail and also required assurance of aircraft and flight testing.

One of the central problems in building IAPs of all types is heat generation by the power section, especially the heat generated continuously due to pump quiescent power losses. Investigation of pumps and discussions with pump manufacturers in the U.S. revealed that extensive development work would be required before an optimum design—a servo-pump type—would be available. This meant that until such time as optimum pump designs became available, IAPs would by necessity use conventional pump designs, with the attendant heat problem and relative inefficiency.

The problem of heat, both generation and dissipation, was thoroughly investigated. A thermal analysis was performed on the simplex and duplex designs. The simplex design successfully passed the high temperature requirement encountered in the F-4 aircraft, without any type of heat exchanger.

Three simplex packages were built by this program. One package was used for flight qualification and was subjected to all the functional and environmental conditions of the F-4 stabilator actuator, including life and endurance tests. Reliability studies were also performed. The two remaining simplex packages were functionally tested with and without load, and delivered to the McDonnell-Douglas Aircraft Corporation for eventual installation and flight test in the longitudinal axis of an F-4 aircraft.

The simplex IAP had a net weight of 88 lb, was of the moving cylinder configuration, and had a steel body (an aluminum body is presently installed in the F-4). Command inputs were via the pilot's manual signal linkage, the stabilization augmentation system (SAS), and autopilot electrical inputs. The system operation is in three distinct modes: manual, SAS, and autopilot. In the manual mode, the electrical unit is inoperative and spring loaded to center. In the SAS mode, inputs from the aircraft motion sensors are summed in series with the pilot's input to position the actuator. SAS signals are converted to mechanical motion by the limited authority auxiliary ram which is a single-channel electrohydraulic servo. In the autopilot mode, the auxiliary ram has full actuator rate and position authority. Provisions are made in the mechanical summer to permit the pilot to override the autopilot by applying sufficient force at the stick.

The emergency hydraulic supply is integrated into simplex. The electrical motor is mounted rigidly to the actuator and contains its own air cooling circuit. The system is sized to provide emergency landing requirements of 10°/second of stabilator rate, and sufficient actuator output force for aircraft control up to Mach 0.9. During conventional operation, the emergency part of the system is inoperative. The emergency system can be initiated at the pilot's discretion, or it comes on automatically and to full power within 1 second if the P_1 hydraulic power is lost (or drops below 500 psi).

The signal conversion system on the duplex consists of quadruple-redundant electromechanical units, which convert electrical command signals directly to a mechanical signal that operates the main control valve.

The results of the R&D program were evidence that the integrated actuator package concept could contribute significantly to improving the survivability of military aircraft when applied to the primary flight control systems, including those equipped with fly-by-wire designs.

Simplex Design and Operation

Guidelines for the simplex design were as follows:

1. It must be interchangeable both physically and functionally with the present longitudinal control actuator of the F-4 aircraft.
2. It must contain a complete backup or emergency hydraulic supply unit supplied by the aircraft electrical system.
3. It must have sufficient hydraulic power to permit recovery on land if main hydraulic supply becomes desirable.
4. It must bring the emergency supply on automatically in event of failure of main supply or at the pilot's discretion.

The actuator barrel is a two-piece unit separated by a center dam, referred to as rip-stop construction. If a crack caused by fatigue or a break due to shell fragment occurs in one section, the propagation of such is prevented from crossing in to the other section. Further, the rip-stop preserves the purpose of a tandem actuator with separate hydraulic supplies in that a single break in the barrel would only affect the loss of fluid in one system.

The dual tandem actuator is made from 4340 steel heat treated to R_c 43–46 (180,000 psi T.S.) and cadmium-plated externally for environmental protection.

The simplex unit hydraulic system had a maximum output flow of 5.5 gpm and pressure of 1600 psi, using hydraulic fluid Mil-H-5606. See Figure 22 for the hydraulic circuit.

Operation of the simplex (Figure 22) is as follows. The aircraft's two hydraulic systems are connected to the package at P_1 and R_1 and at P_2 and R_2. Filter screens F_1 and F_2 prevent large contamination particles from entering the system. Check valves are installed in the two inlet lines to prevent backflow from the package in the event of a ruptured pressure line upstream of the package. The aircraft's Hydraulic System 2 powers the aft lug end of the actuator, and the aircraft's Hydraulic System 1 powers the forward rod end of the actuator, both being controlled by the dual tandem main servovalve. System 1 also supplies the hydraulic power for the auxiliary actuator, including the electrohydraulic servovalve, auxiliary ram, unlocking pistons, authority stops, and input linkage locking pistons. This emergency system is designed so that it can be brought on-line in place of P_1 to power the actuator's rod end. Position of the switching valve dictates whether P_1 or the emergency system is on-line. The emergency system reservoir is continuously filled by the return pressure of System 1 through an orifice and check valve arrangement that eliminates the need for external filling of the system.

In addition to qualification and environmental tests on the simplex package, both survivability and reliability analyses were conducted.

Duplex Integrated Actuator Package

The duplex IAP (Figure 23) developed on this program consists of a two-piece dual tandem steel actuator controlled by a redundant fly-by-wire signal converter and powered by dual self-contained hydraulic power supplies and associated hydraulic components. The performance requirements were based on the requirements of the F-4 stabilator actuator. Input commands to the duplex actuator were quadruple, requiring a signal converter that was to be dual fail-operative (i.e., could operate with a small degraded performance after two failures).

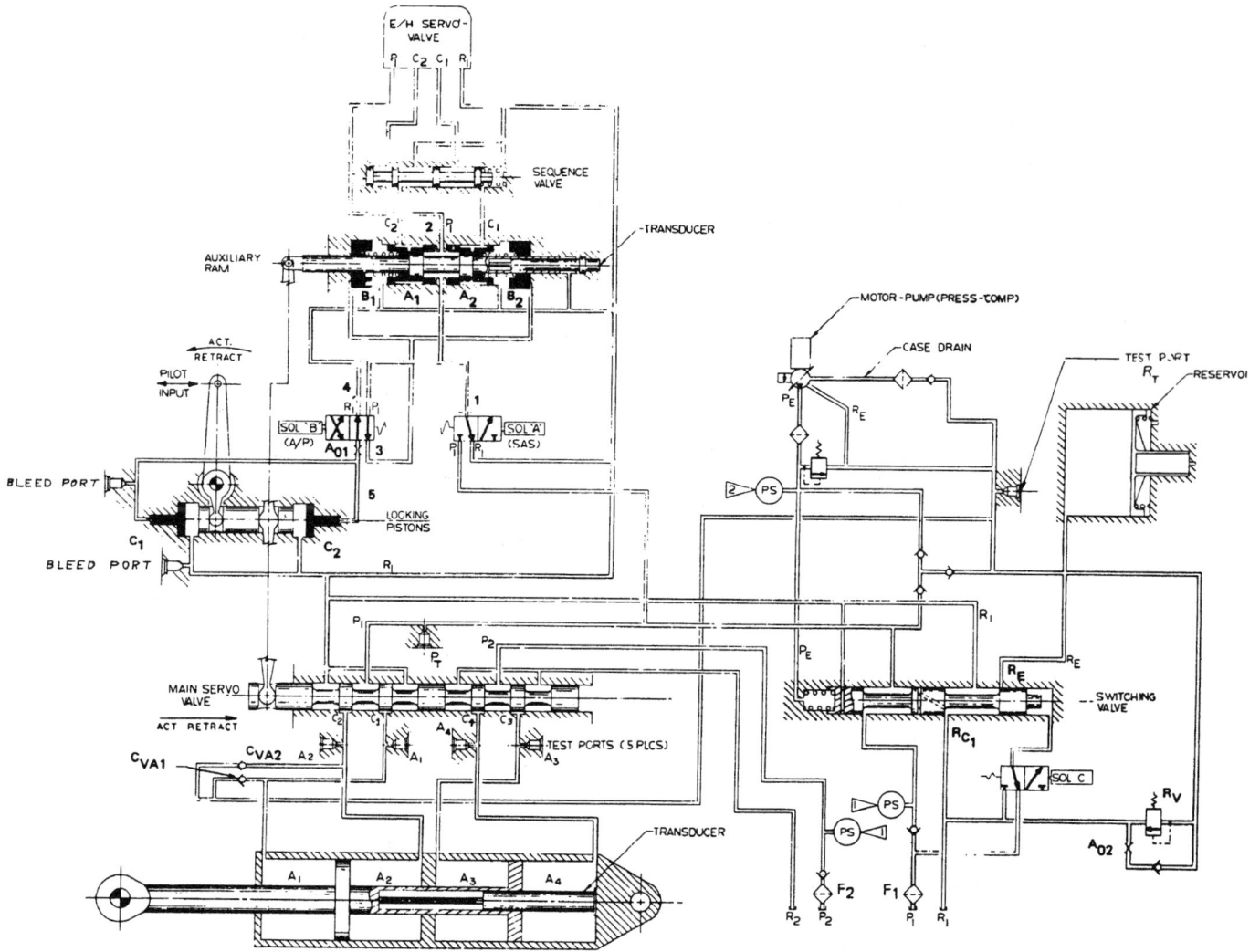

Fig. 22   Simplex hydraulic circuit.

Fig. 23   Duplex actuator.

The duplex signal converter consists of four electromechanical actuators. The outputs are force-summed on a torque tube that in turn operates the dual tandem servovalve for controlling hydraulic flow to the actuator (see Figure 24).

The duplex IAP operation is as follows. The electromechanical actuators drive the main valve simultaneously. The output of each actuator is transmitted to the main servovalve through a funk spring. Funk spring breakout force is equal for all actuators but is higher in one direction than the other to permit positive control under all conditions. Should any two of the actuators fail in any combination or sequence of jams or opens, the remaining two have the capability of driving the main valve by breaking out the funk springs of the failed units. The difference in force required to break out the funk spring in one direction as opposed to the other ensures that at least one of the remaining units is operating within its detent range. A transient effect is produced in reaching the new position where at least one spring will be in its detent range after a failure; however, the system retains its stiffness no matter what the failure.

Laboratory Tests

The duplex package was given functional tests under standard conditions, including frequency response, failure transients, slew rates, threshold, and load. These test results showed that the package met the stated design and performance requirements.

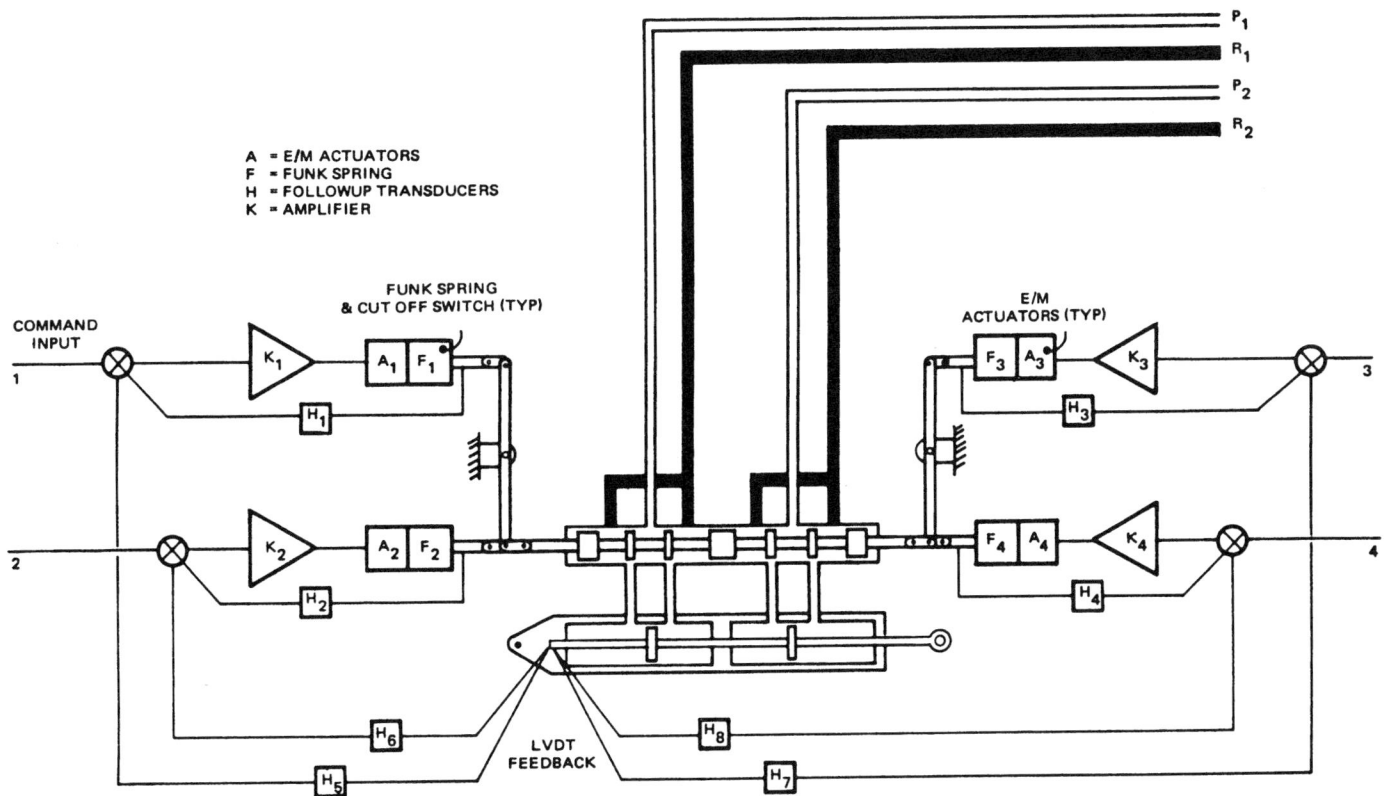

Fig. 24   Force summing with four electromechanical actuators inside the feedback loop.

Comments on the LTV Program

As the LTV program was originally planned, it would have produced a laboratory model that solved or overcame the technical problems in building an actuator package qualified for use in flight control design, similar to what the British had accomplished on their VC-10 aircraft. However, small arms and ground fire then occurring in the conflict in Southeast Asia (the Vietnam War) required a design aid to improve or increase the survivability of military aircraft to a hostile ground environment, so the effort was redirected to fill this military need. The revised program necessitated the use of state-of-the-art components, and allowed practically no time for extended research on the really tough problems involved, such as the heat rejection problem. This program produced an integrated actuator package (IAP) called the simplex.

The IAP system cannot eliminate the excess heat through a heat exchanger, as is standard in the conventional design. Conventional aircraft design has an advantage over the IAP concept with regard to heat generation in that the IAP either must find a method to eliminate the heat it generates or must live in the heat environment. The conventional design has long lines running from the pump (one heat source) to the actuator (the second heat source), hence the lines can release the excess heat to the surrounding air. In addition, a heat exchanger is installed in the fuel tanks to reduce the hydraulic fluid temperature, the fuel being consumed by the engine(s).

This program was quite successful in that most of the effort was directed toward application of a concept rather than producing an individual design or a design with many uses. The purpose of the redirected program was merely to provide a design that had a "get home and land" capability; it was not intended as a replacement for any current design. The simplex filled a one-hydraulic system supply; and as the aircraft had two hydraulic systems, the next step would be to design a duplex package.

Further, it was envisioned that all actuator packages used in flight control designs would be electrically controlled, hence the use of fly-by-wire. Therefore, it was simple logic that the most survivable design would be IAPs with fly-by-wire, thus planting the seeds for Advanced System Development.

Sperry Phoenix Programs

The work by Sperry is divided here into two parts primarily because it is a useful method of organizing the program's events and data. Essentially, Part I covers the results of work performed on the initial contract. The results of this effort are reported in AFFDL-TR-67-53, the source of the data presented in this discussion.

Sperry Phoenix Part I

Sperry's initial effort was started in 1966. A follow-on effort the next year was eventually completed in 1968. The initial work was basically to establish the system design requirements and trade-offs, as well as the requirements of the types of components to be used, control signal format, method of transmitting signals, actuator configurations, degrees of redundancy, failure detection techniques, and artificial feel mechanization.

By definition, a fly-by-wire flight control system is an electrical primary flight control system employing feedback such that vehicle motion is the controlled parameter. No mechanical backup is used. Before 1966, fly-by-wire had been studied and proposed for at least 20 years, often under the title "Electrical Flight Control Systems." However, past research had nearly always been narrowly aimed at one or two specific approaches to replace the link between the control stick and the surface, ignoring the handling quality or feel requirements. The Sperry effort used a more general approach.

Although mechanical control system designs had improved tremendously over the years, both in techniques and materials, they had a progressively difficult time keeping up with the performance gains and control requirements of successive generations of aircraft. Most designers had agreed that fly-by-wire could solve the flight control problems if a practical approach could only be mechanized. The problem had been that no one provided a practicable and reliable fly-by-wire system design that could be produced with existing hardware. This problem had several facets. One primary factor had been the unavailability of components having proven reliability. Another factor is that a fly-by-wire design is a multidisciplined venture that encompasses mechanical, electrical, hydraulic, and some aerodynamic engineering. Further, the application of redundancy was not generally well understood. The Sperry effort proposed solutions to these factors and demonstrated how a practicable redundant fly-by-wire system could be mechanized using available hardware.

Sperry personnel made a number of visits to other organizations in the aerospace industry to obtain data on fly-by-wire designs. Visits were made to the following companies:

Grumman Aircraft Company
The Boeing Company
North American Aviation
Douglas Aircraft
General Dynamics
Hydraulics Research and Manufacturing Co.
National Water Lift Company
Weston Hydraulics Ltd.

Because reliability is such an important factor in design of fly-by-wire systems, particularly because the number of channels required is directly related to the reliability of the various components used, Sperry discovered very early in their program that many times the reliability numbers became a "numbers game." This became very apparent in the electohydraulic and electromechanical designs, where correlations of data between simple designs and the more recently applied complex designs were practically nonexistent.

Flight Control Design versus Time

Flight control designs change with changes in requirements, and requirements change with time. The following figures and descriptions demonstrate the various designs and why the changes were made.

The simplified control system shown in Figure 25A is a simple reversible system still used in light aircraft. Here, all control forces are reflected back to the pilot's hand.

In the irreversible fully powered system shown in Fig. 25B, an artificial force producer must be added for the pilot to feel that aerodynamic forces are acting. Control of the neutral position of the feel system as it changes with flight condition is also required. This is a trim function very similar to the simple system.

A parallel input servo (Fig. 25C) moves the stick along with the pilot. Such a servo commonly provides AFCS (automatic flight control system) inputs so that the pilot can observe and monitor its actions. The series input servo

A

B

C

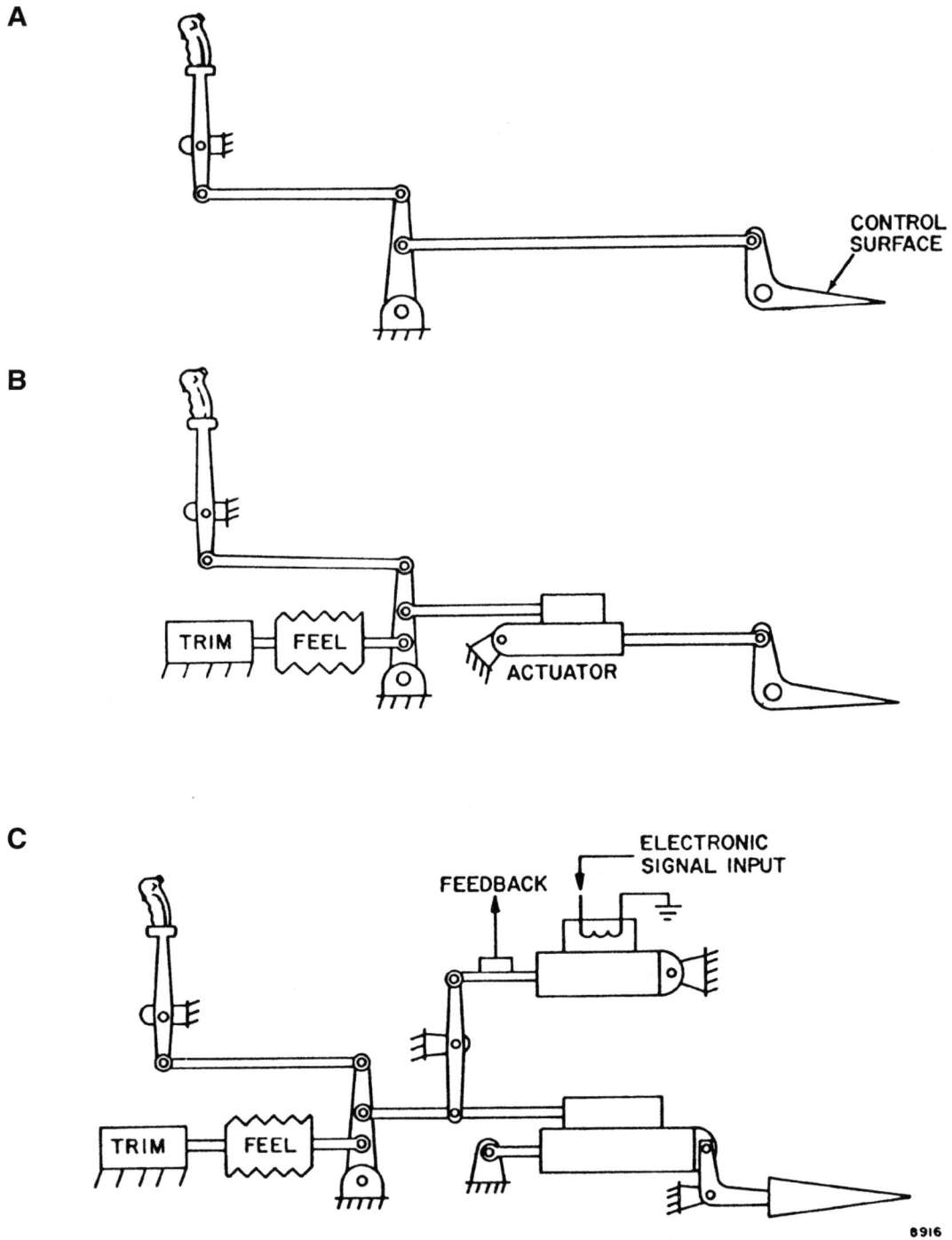

Fig. 25 *(A) Simplified aircraft control system. (B) Fully powered (irreversible) control system.*
(C) Parallel input servo.

(Figure 26A) adds to or subtracts from the pilot's inputs so that no control stick motion occurs. This type of servo is employed for stability augmentation.

Fig. 26B shows another method for adding series inputs that makes for a lighter control system. Figures 26C and 27A show two types of control stick steering mechanization. These designs are also called command or control

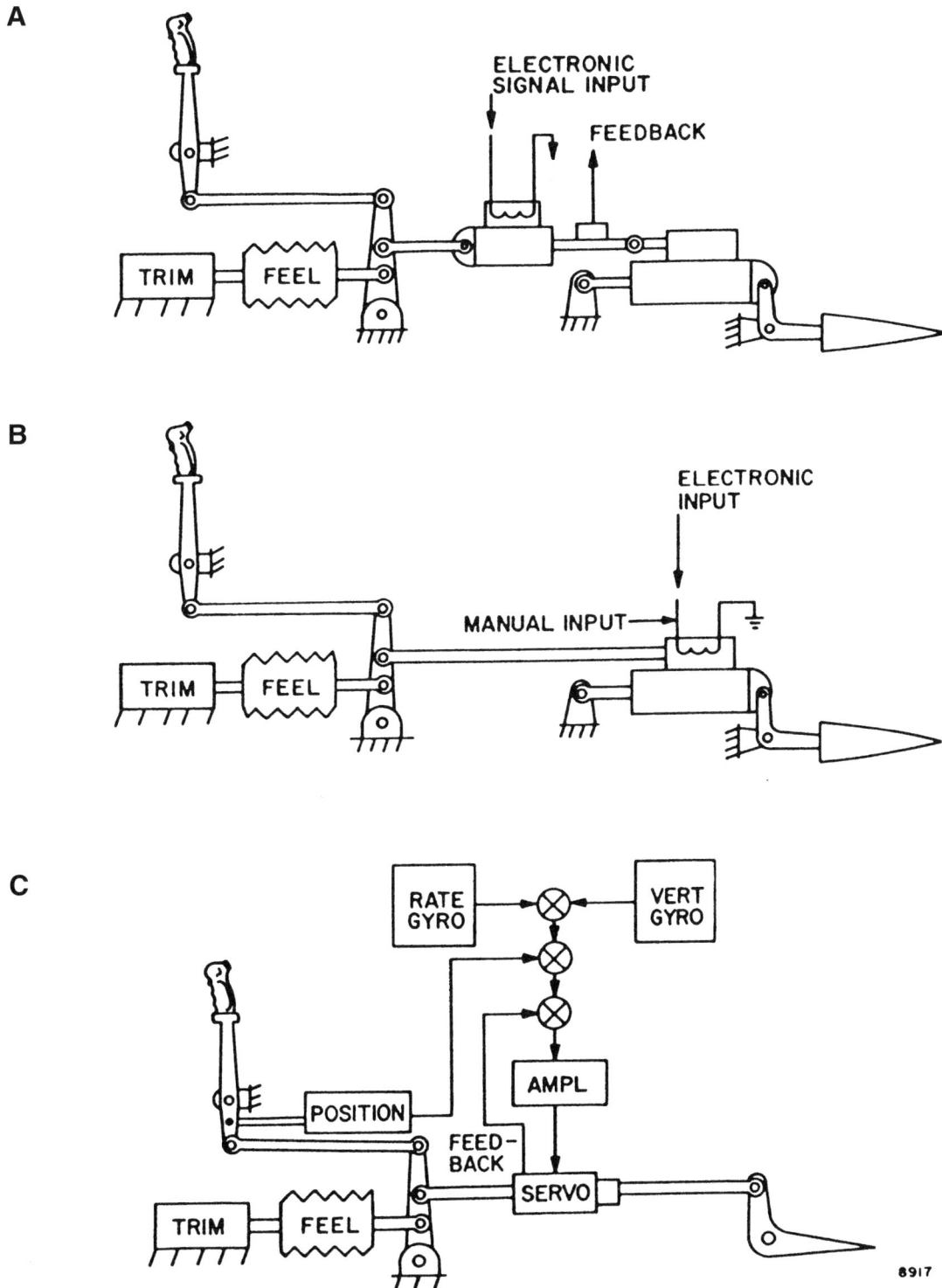

Fig. 26 *(A) Series input servo. (B) Dual valve input. (C) Differential servo type of control stick steering.*

augmentation systems. Their purpose is to improve control response by bypassing control system friction, inertia, deadzones, or any other problem in that particular system. Control stick steering or command augmentation is used on the F-111, the A7A, the supersonic transport (SST), and the jumbo jets (e.g., the C5A), and is the forerunner of the fly-by-wire control system shown in Fig. 27B.

39

A

B

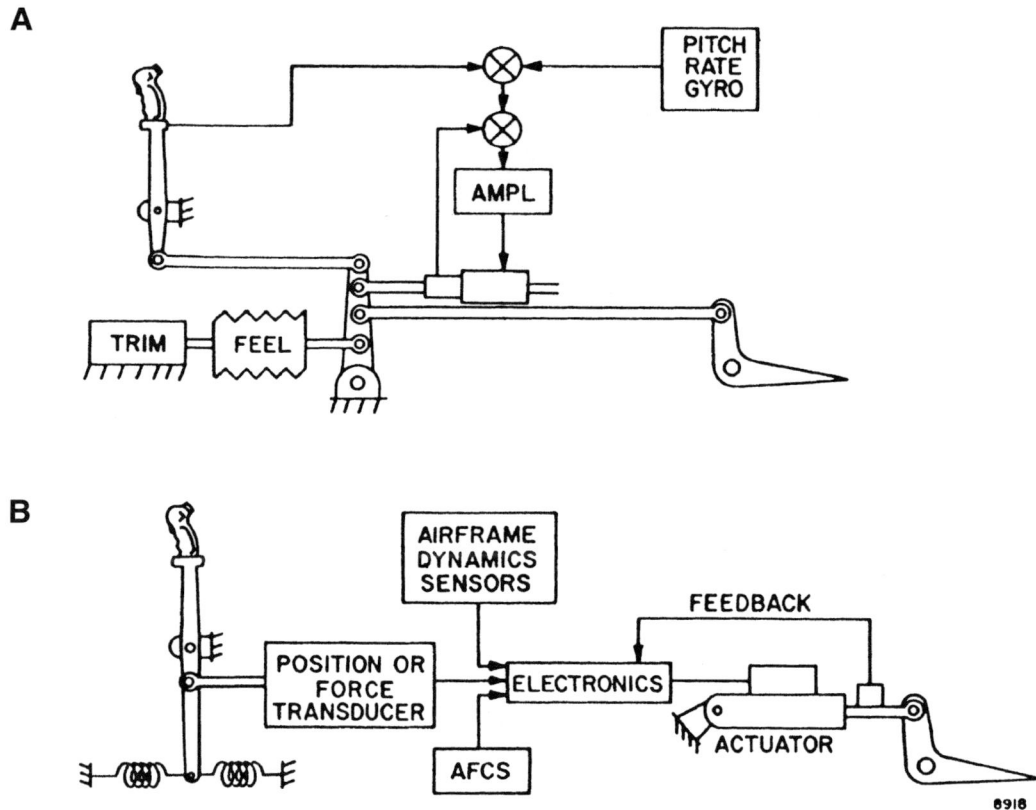

Fig. 27 (A) Force stick type of control stick. (B) Fly-by-wire control system.

Use of Redundancy

The application of redundancy to flight control design is of particular interest in fly-by-wire technology for the required system reliability and safety issues. Redundancy has been to aircraft flight safety what systems such as SAS, CAS, or AFSC have been to all-weather landing (AWL) mode. Triple redundancy with voting is a brute force technique—it is inefficient and adds undue complexity, cost, and weight to the system.

Fly-by-wire systems require a greater failure tolerance as they must operate after double failures to obtain the desired degree of reliability and safety. Unless a fly-by-wire design employs redundancy, its reliability will never meet or exceed that of a mechanical system. Sperry Phoenix had had a program for several years that used finesse in optimizing redundancy. The technique, called *fail-passive design,* designs out the cause of active (i.e., hardover) failures so that the resulting channels or components fail in a passive manner only. A fail-passive component or channel fails in such a way that it has no output and does not interfere with the normal operation of a parallel component or channel. By using fail-passive design, a fail-safe system requires only one channel, not two; a fail-operational system requires only two channels, not three. Furthermore, little or no monitoring or switching equipment is necessary. A system using this design technique represents three orders of magnitude of improvement in system reliability. The reliability criterion, or rather the probability of failure criterion used on this program, has been established as 2.3×10^{-7} each hour.

Actuator Redundancy

The design of actuators that will operate after double failures involves a large number of parameters and trade-off factors. Two important design factors are 1) failure rate, including false-alarm rate, and 2) switching time and transients caused by a failure. The maximum switching time criterion is based on the maximum allowed normal acceleration or displacement before the flight condition for the aircraft in question is regarded as unsafe. The time is measured from the onset of a hardover failure to the restoration of normal operation, including the time to detect the failure and the time to switch out the failed channel. A time of 50 milliseconds was assumed on this program.

Actuator design trade-off factors involved include:

1. degree of redundancy (dual, triple, quadruple)
2. type of redundancy (active or standby)
3. position or force summation
4. type of servovalve (flow or pressure control)
5. hydraulic or electronic monitoring
6. whether secondary actuator is used
7. mechanical or electrical feedback

Design Comparison

Fly-by-wire control has a number of general advantages over the conventional mechanical designs:

- Improved control performance through better dynamic response and the elimination of friction, backlash, hysteresis, compliance, and inertia of the input controller.
- Smaller installed weight and volume.
- Reduced costs of logistics and maintenance.
- Better logistics and maintainability because of the reduction in number of critical parts, easier access, and higher level of interchangeability between aircraft.
- Reduced vulnerability to minor structural damage, maintenance errors, or a hostile environment (battle damage).
- Greater flexibility in cockpit installation and orientation.
- No coupling into body bending modes.
- Reduced required design effort.
- More flexibility to design or change performance.

In addition to eliminating all the push rods and bellcranks of conventional designs, fly-by-wire permits the use of side-stick controllers, allowing the displays to be moved closer to the pilot and reducing pilot-inertial coupling.

The following two illustrations permit a comparison between a conventional design and fly-by-wire. Figure 28 is a simplified diagram of the F-111 mechanical pitch/roll control. Figure 29 shows an equivalent fly-by-wire system mechanization of the F-111 design.

Performancewise, a desirable system is best obtained through a closed-loop approach in which the desired flight path is the input and the actual flight path and dynamic response parameters to be controlled are fed back. One such

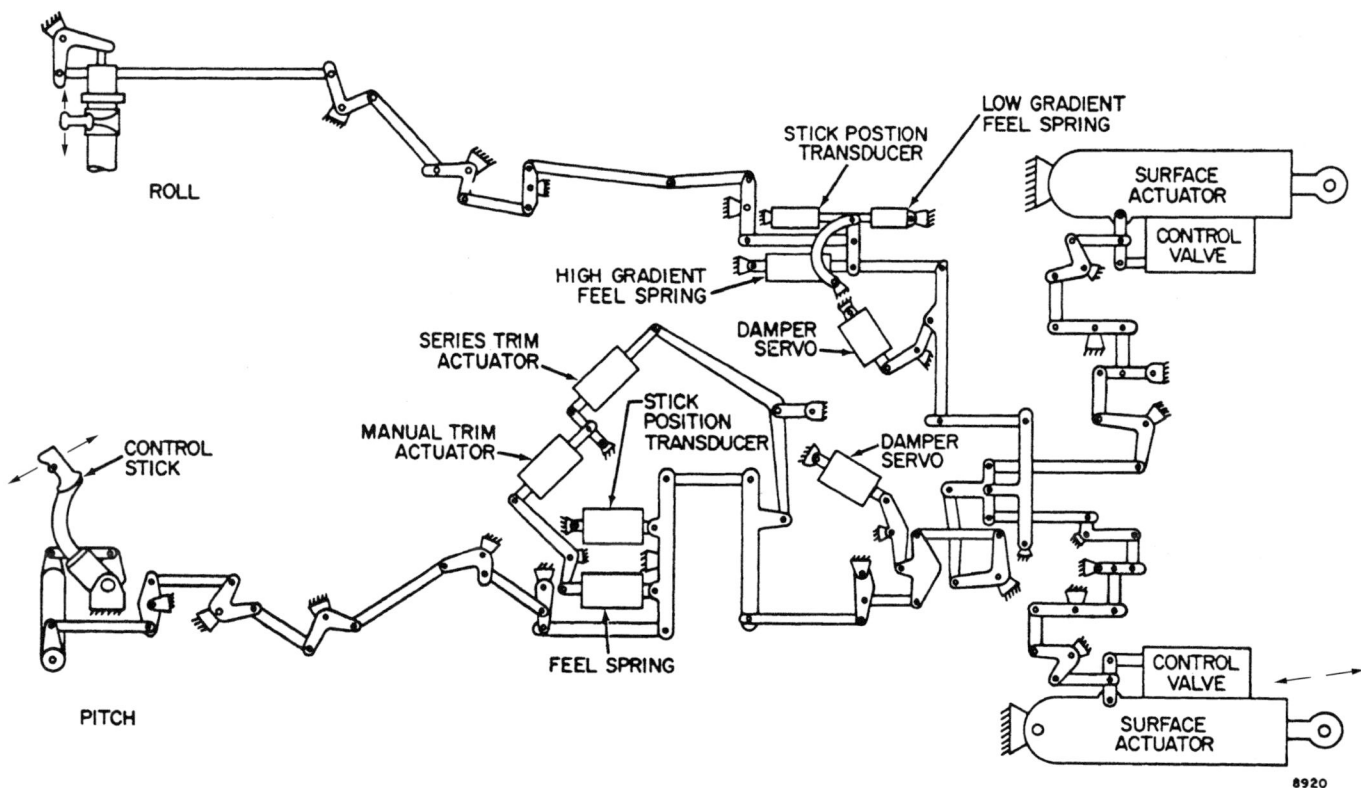

Fig. 28   F-111 pitch/roll mechanical control system.

design, C*, employs a blend of pitch rate and normal acceleration feedback in the pitch axis. Artificial feel implemented by the C* command approach has a number of advantages over other methods:

- It provides nearly neutral speed stability which permits tracking during rapid speed changes without trim.
- Aircraft response conforms to angular rate at low speeds and normal acceleration at high speeds.
- The system is independent of airspeed or altitude.
- It provides good command response while maintaining high gust damping.
- It is flexible; for example, signals from other subsystems can be added.
- Feel is independent of the type of aircraft.
- Sensors used in most aircraft to augment stability could also be used.

Because of its relative simplicity and more natural feel characteristics, the C* command is recommended for use in the basic fly-by-wire system.

Actuator Configurations

The Sperry Phoenix program evaluated seven actuator design configurations for fly-by-wire systems, but only two are discussed here to provide a comparison between designs.

The configuration shown in Figure 30 is a conventional standby-redundant actuator with electrical monitoring on valve spool position. It uses the most familiar concepts, has three real channels, and has a model channel that may be

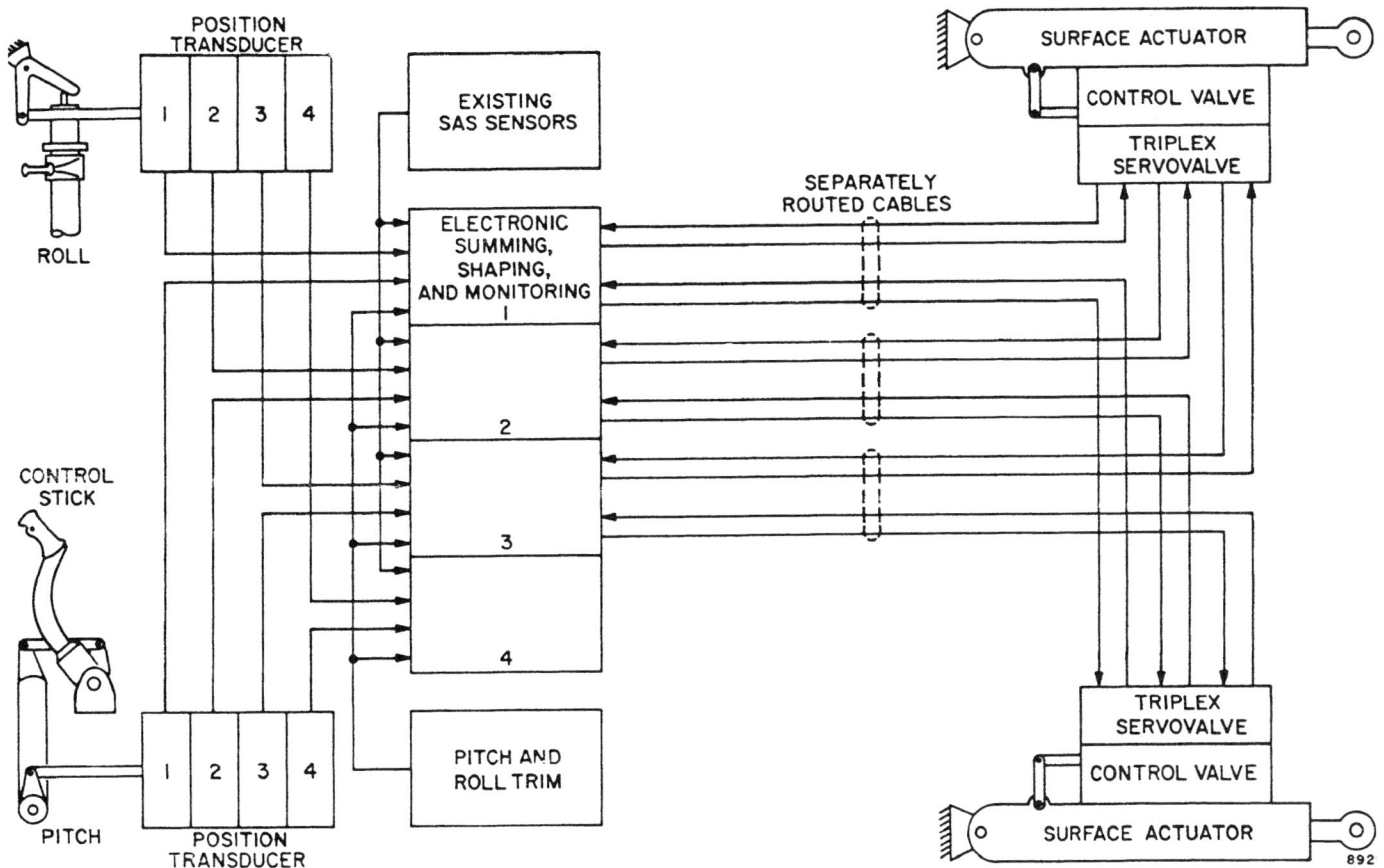

Fig. 29   Equivalent fly-by-wire system.

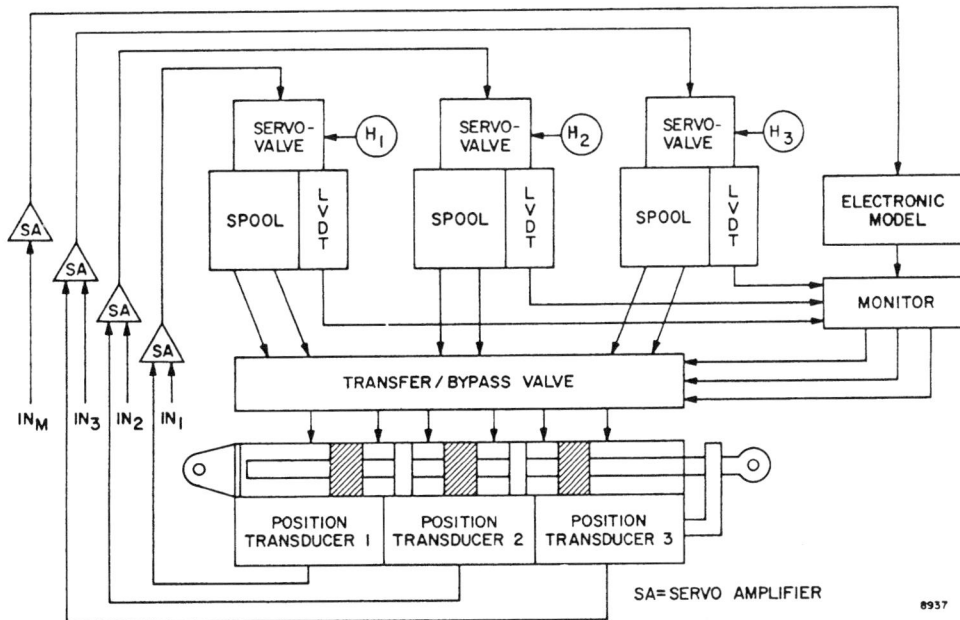

Fig. 30   Conventional standby-redundant actuator with electrical monitoring.

43

hydraulic or electronic. One real channel is active while the other two operate in standby mode. The two-stage servovalves are coupled to the actuator through a four-position engage valve which transfers the system through its operational modes on commands from the electronic monitor via electrohydraulic solenoids. Position transducers on the servovalve spools and model channel provide signals for comparison monitoring. This design has several disadvantages:

1. Fast transfer times require high-speed solenoids and comparators to minimize transients.
2. The size and weight increase, as each actuator must be sized to carry the full load.

The configuration recommended for fly-by-wire systems is the fail-passive secondary actuator design, shown in Figures 31 and 32. A small redundant secondary actuator is used to mechanically drive the main control valve and power actuator. The secondary and power actuators employ active redundancy. If dual hydraulic supplies are used, the secondary actuator is dual tandem, with two single-stage jet-pipe servovalves driving each piston, thus forming four inner servo loops. This configuration derives its uniqueness from the inner loops, which are designed to have passive-fail characteristics. A fail-passive channel fails in such a way that it has no output and does not interfere with the normal operation of a parallel channel, thus eliminating active or hardover failures. As a failed channel has no force output, the other good channels can operate unimpeded. The electronics fail passively because AC signals are used; a hardover electronic failure causes a DC output to which the AC circuits are insensitive. Also, should a hardover input occur in a channel, the other channels collectively offset the output force of the failed channel at the force-summing actuators. The high loop gains reduce the resulting position offset to an insignificant level. Further, no monitoring, switching, or engage valves are required on this design. The fail-passive triplex and quadruplex designs were tested in the laboratory to demonstrate and validate the fail-power secondary actuator design technique.

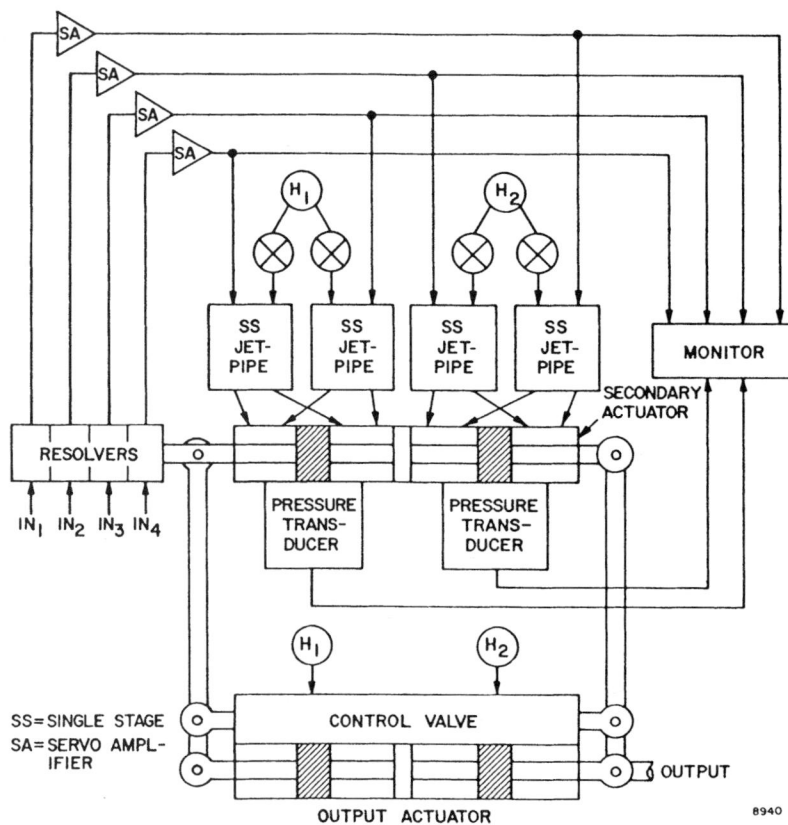

Fig. 31   Fail-passive secondary actuator design (I).

44

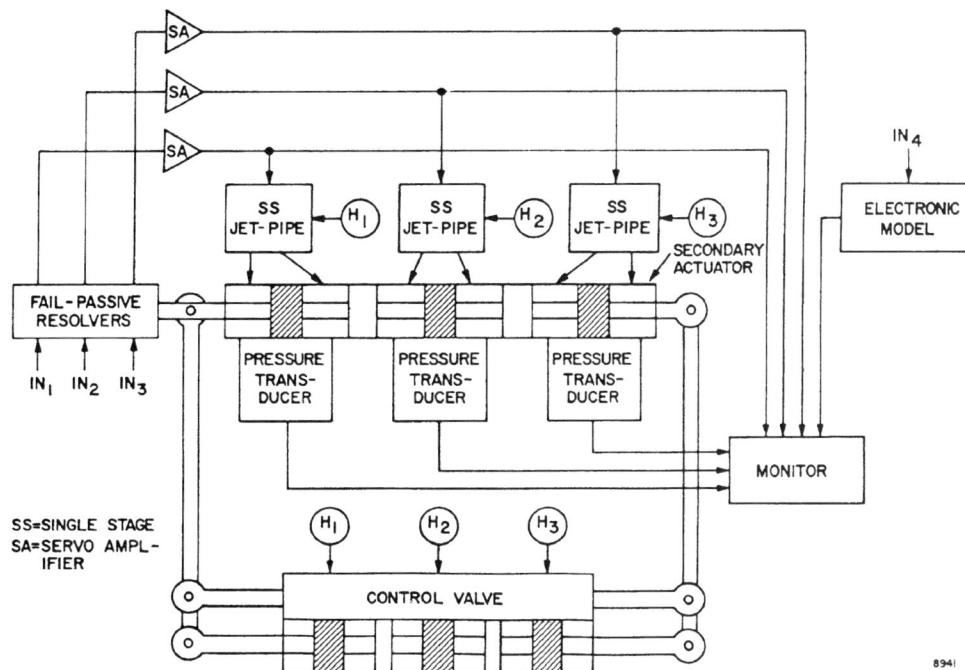

Fig. 32   Fail-passive secondary actuator design (II).

Program Milestones

Years of experience in building thousands of aircraft autopilots with electronic components gave Sperry the unprecedented technical background that was required to develop the design techniques to apply fly-by-wire with greater confidence. The needed elements that Sperry researched were:

- Establishing the redundancy necessary for fly-by-wire systems to have reliability and safety equal to or better than the mechanical designs.
- Devising design methods of detecting and removing failures in a system and doing so without producing undesirable transients.

Devising the monitoring, logic, and voting schemes to support redundancy and failure detection was Sperry's major achievement. It proved at last that now we could do the job electrically, instead of mechanically.

A laboratory experimental model of a fly-by-wire system was constructed and used to evaluate and validate the program's design techniques. The Sperry program also made significant contributions to the technology of artificial feel systems and in the area of flight control actuator redundancy.

Sperry Phoenix Part II

Program Objectives

The Sperry program had three primary objectives:

1. To develop fly-by-wire technology and design criteria for manned aircraft.
2. To employ state-of-the-art hardware in the construction and evaluation of an experimental laboratory model fly-by-wire system;
3. To apply fly-by-wire design techniques to a specific aircraft—the B-47 used in in-house programs.

Laboratory Model

The Sperry design was a two-fail operational system composed of channel electronics, control simulator, side-arm controller, and triple-redundant electrohydraulic actuator.

B-47 Aircraft Analysis and Control Simulation

Sperry conducted an analysis of the B-47 longitudinal axis with the C* feel system during various flight conditions. Figure 33 is a block diagram of the longitudinal axis. Figure 34 shows the simulation results of response to three flight conditions:

I. Altitude 10,000 ft, true airspeed 174.3 mph
II. Altitude 35,000 ft, true airspeed 476 mph
III. Altitude 20,000 ft, true airspeed 505 mph

The normalized C* traces essentially follow the pilot forward path command prefilter, and thus remain within the desired C* envelope.

The control simulator consisted of two main components, the simulator and the side-stick controller. These are shown in Figures 35 and 36. The simulator provides aircraft attitude information in response to control stick inputs and allows the pilot to simulate various system failures at any time and observe their effect on the simulated aircraft. Manual input commands of the control stick are converted into aircraft motions that are displayed by the flight director and rate meters on the panel of the control simulator. The simulator contains a pilot's control panel which permits one to engage and select the various modes of operation of the system. The simulator also has a failure display panel to indicate the operational status of the fly-by-wire system by axis. The simulator has a preflight self-test system that permits a preflight self-test, the results of which are shown by the go-no-go lights. When coupled with the side-arm controller, actuator, channel electronics, and necessary power supplies, the control simulator made a complete demonstration unit.

Channel Electronics

A considerable amount of electronic circuitry was used on this program, of which Figure 37 is a representative example. The fly-by-wire electronics were separated into four channels: three identical active channels and a model channel. Each of the active channels drives one section of the triplex actuator, and the model channel has an actuator pressure model.

Active Channel Electronics

• **Computation**   The electronics were implemented using basic DC computation techniques. The system is DC with the exception of the demodulators on the stick position rotary variable differential transformer (RVDT), and the differential pressure transducer inputs. Full-wave demodulators were used in all cases. The advantages of full-wave demodulation are higher bandwidth, better noise rejection, and more symmetrical gain characteristics.

• **Integrator**   The forward loop computation consisted of an integral plus displacement computation that is accomplished in a single amplifier. The integrator synchronization during the disengage mode is accomplished by feeding back the sum of the synchronizer output (servo command) and the RVDT output. Each integrator equalization is

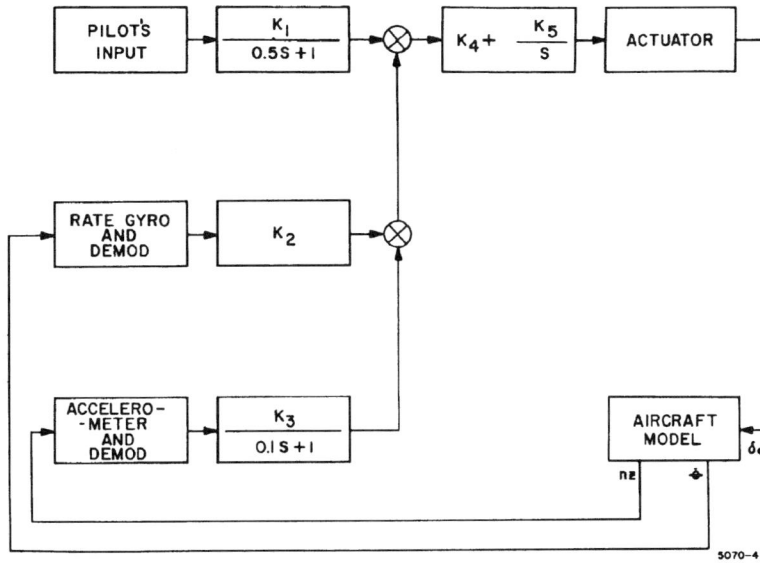

Fig. 33   Longitudinal axis block diagram.

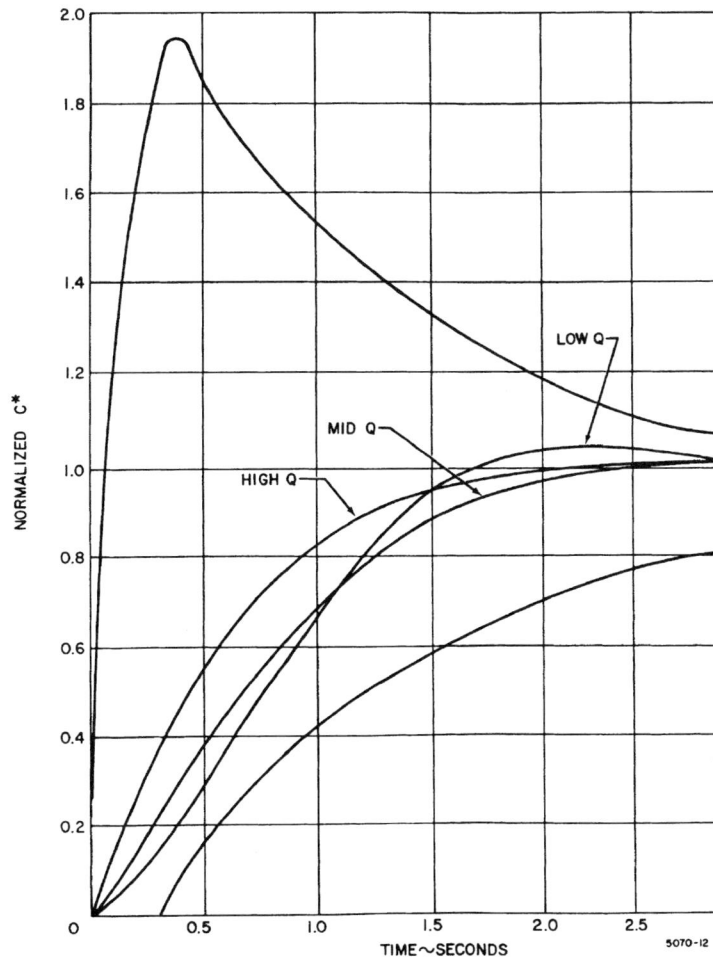

Fig. 34   Normalized C* responses at flight conditions I, II, and III.

47

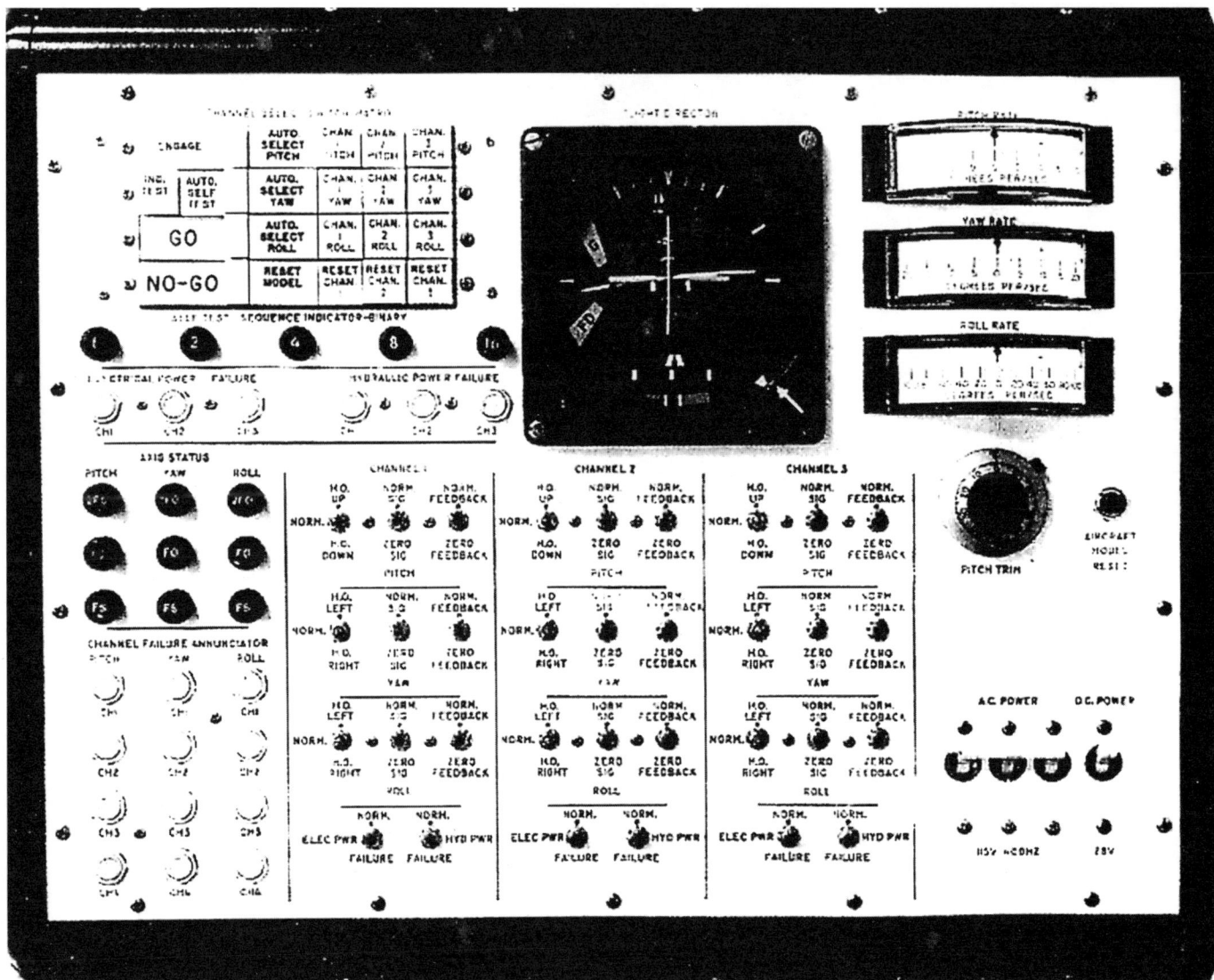

Fig. 35   Control simulator front panel.

accomplished by feeding back the difference between the midvalue of the three voted servo commands and the output of that integrator. The midvalue is selected using an operational amplifier midvalue logic (MVL) circuit. A 10% unbalance in inputs will cause the feedback amplifier to saturate, thus eliminating the tracer signal and causing the channel failure logic to indicate a failure.

• **Midvalue Logic**   The MVL amplifier operated with no more than three operating amplifiers at any one time. The nonoperating amplifiers have their power removed, in which condition the inputs and outputs all exhibit high impedance. Under a two-fail operational status (no failures), the three active channel servo commands are voted. Upon first failure, unless the first failure occurs in the model channel, the power to the failed channel operational amplifier is removed and the power to the model channel amplifier is applied.

• **Failure Logic**   The failure logic is a self-latching logic loop that requires a number of inputs to remain latched in. The absence of any of these inputs will cause the system to indicate a failure. The logic circuitry does have a

Fig. 36   Side-arm controller.

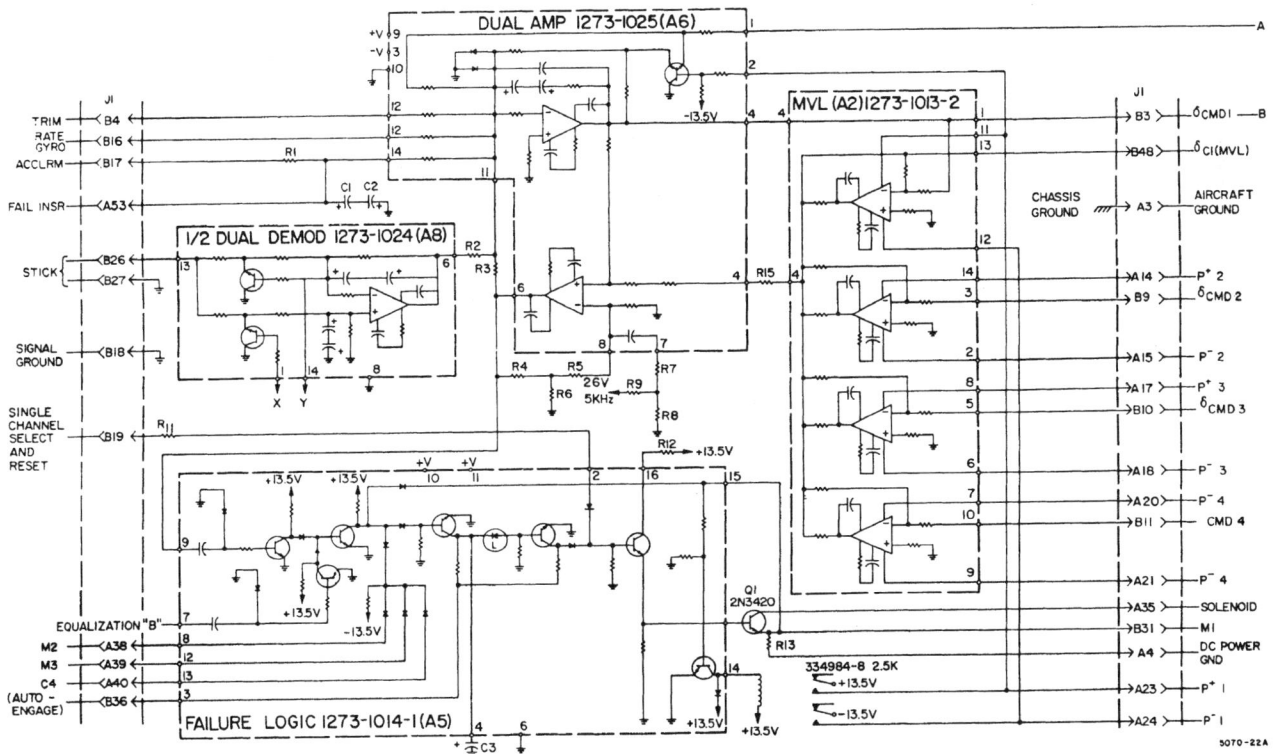

Fig. 37   Schematic of fly-by-wire model channel electronics.

49

200-millisecond dropout delay to protect against nuisance trips. Any failure lasting fewer than 200 milliseconds will not result in a failure indication.

• **Servoamplifier**   The servoamplifier uses current feedback such that valve inductance phase lags are reduced to a negligible value. The pressure equalization signal is the difference between the pressure output of that channel and the midvalue of the voted pressure signals.

Triple-Redundant Actuator

The prototype triple-redundant actuator in Figures 38 and 39 was developed to demonstrate the actuator concept for a fly-by-wire flight control system. The actuator consists of three electrohydraulic servomechanisms arranged in a side-by-side manner. By the separate housing approach, intersystem leakage and crack propagation are eliminated. The three systems may be operated individually or simultaneously. Figure 40 is a schematic diagram of one channel.

A command signal from the servoamplifier to the servovalve causes fluid flow to the servo ram. Position feedback from the servo ram is provided by a rotary variable differential transformer (RVDT). The feedback signal is summed with the command signal and the servo ram moves until the sum of the signals is equivalent to the servovalve current needed to balance the spring load on the servo ram. The load on the servo ram includes the force output from the other two servo rams in a triplex system. If a discrepancy among the three systems occurs, a force fight among the

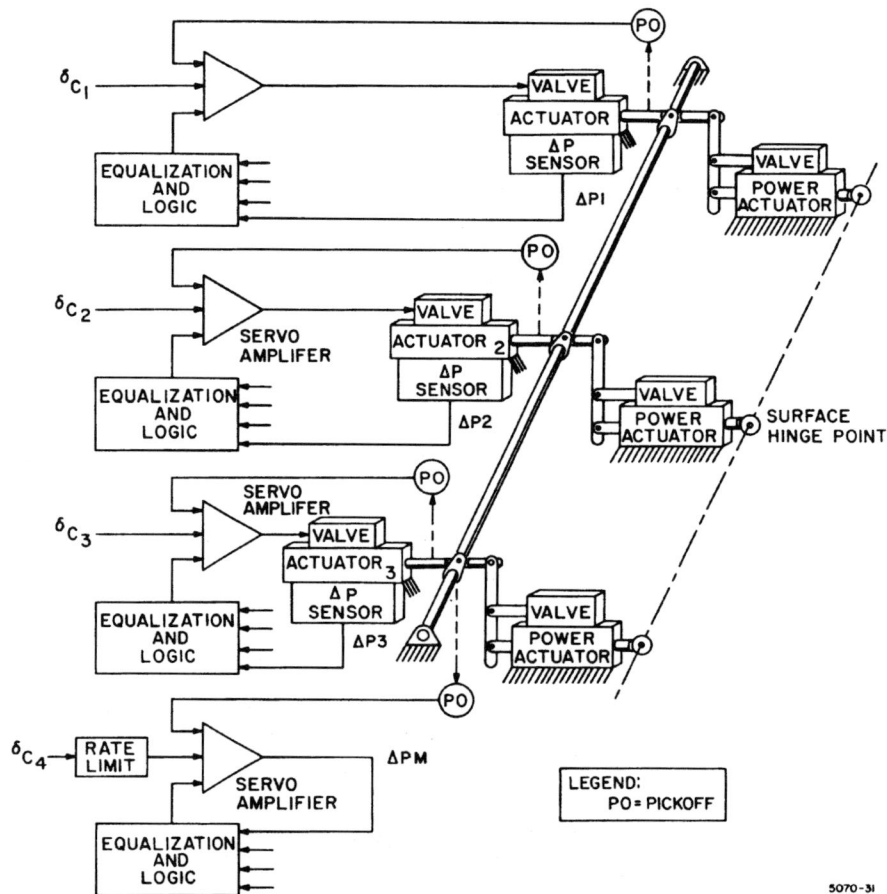

Fig. 38   Redundant actuator concept.

systems results. To reduce the force fight to a minimum, pressure equalization is employed. Differential pressure transducers measure the pressure output of the individual channels. These signals are voted in a midvalue selector and an error signal is generated to force the individual channel to approach the midvalue.

The triple-redundant actuator was given a series of laboratory tests which showed that it met all the design and performance requirements. However, it was never given flight tests.

Fig. 39   Triple-redundant actuator.

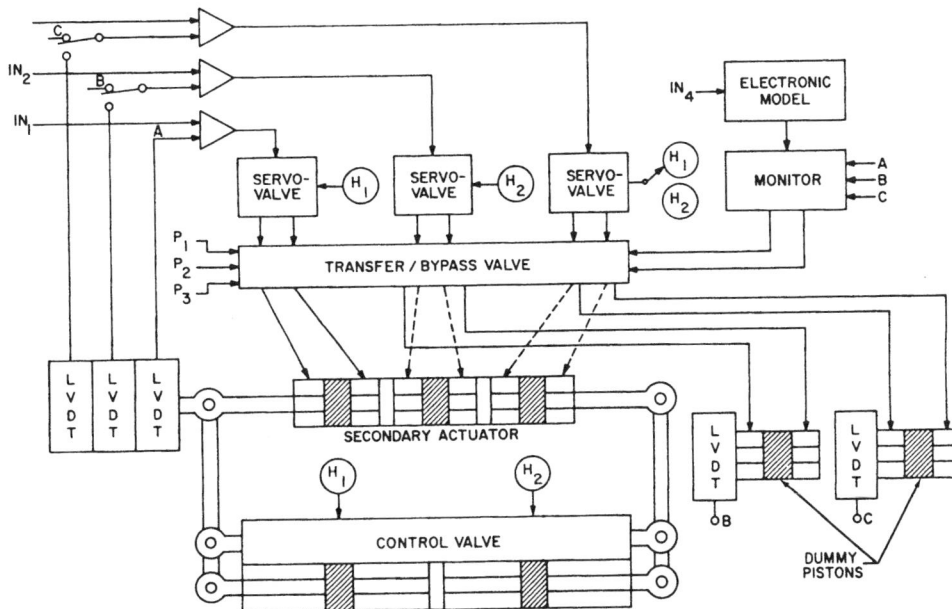

Fig. 40   Triple-redundant pitch channel.

Reliability and Failure Modes

A reliability and failure mode analysis was performed that included the single channel and system of the longitudinal axis. This analysis provided data on the reliability of components and systems currently in use, revealing those items that needed improved performance.

Summary of the Sperry Effort

It was obvious to the persons who directed the Sperry programs that they had established the electronic designs and design technique necessary to make fly-by-wire a reality. All that remained was to apply these designs to real systems. A year after the Sperry program ended, MCAIR contracted with Sperry to provide the electronics for their fly-by-wire system on an F-4 aircraft.

Chapter 4

The Survivable Flight Control System Program

Background

As the name "Survivable Flight Control System" indicates, the system brought together technologies that, when properly implemented, would permit the design of a more survivable flight control system in military aircraft when exposed to a hostile environment. Conceptually, the Survivable Flight Control System (SFCS) Program came into being in the latter part of 1968, while the efforts that established its basic building blocks were winding down. The required preliminary work had been successfully completed with positive results on the Douglas Long Beach fly-by-wire program, the B-47 in-house fly-by-wire program, the LTV effort on integrated actuator packages, and the Sperry Phoenix effort on the electronics for fly-by-wire systems. Also, to accelerate the work on integrated actuator packages and also as a backup to the LTV effort, an additional program had been undertaken with General Electric Co. in Binghamton, New York. Because it started later, the GE work lagged the LTV effort; however, GE did produce a simplex unit, which was laboratory tested but never flight tested. The LTV and GE programs encountered the same problems in designing and building an IAP.

While making plans for an advanced development program, personnel of the Flight Dynamics Laboratory made visits to most of the airframe manufacturers to describe the progress made on fly-by-wire programs, determine the feeling industry had on fly-by-wire systems, and obtain any suggestions flight control system designers might offer to be included in an advanced development program. These visits proved quite worthwhile.

At the same time the contractor, McDonnell Aircraft Co. of St. Louis, Missouri, was organizing the workgroup and making the proposal for the Advanced Development Program (ADP), the Flight Dynamics Laboratory organized a team from various branches and designated them as members of the 680J ADP project office. The initial contract for the program, negotiated in June 1969, was scheduled for completion in 36 months at a cost of approximately $16.5 million.

Constraints

The new 680J ADP team had to comply with several constraints. First, the program's focus was on application only—the research was over. For example, they could no longer work on the IAP's heat rejection problem, but rather had to go on what they had. Also, little could be done to the IAP's weight and size; the aircraft might be potentially altered for IAP installation, but making changes to the IAP's dimensions was no longer possible.

The second major constraint was one aircraft, one program. The entire program was allotted only one aircraft, so each phase of the effort depended on the completion of the preceding one. Had at least two aircraft been given to the program, any modifications could have been made at the same time; with only one aircraft, modifications required substantial down time. Also, if the only available aircraft was lost for any reason, there was a reasonable doubt that the program would ever be completed.

Program Phases

The ADP program was conducted in the following phases:

Phase I	Simplex Package Evaluation
Phase IIA	Fly-by-Wire with Mechanical Backup
Phase IIB	Fly-by-Wire without Mechanical Backup
Phase IIC	Fly-by-Wire with Survivable Stabilator Actuator Package in Pitch Axis
Phase IID	Demonstration Flights
Phase IIE	Precision Aircraft Control Technology and Reliability/Maintainability of SFCS Fly-by-Wire

Phase I: Simplex Package Evaluation

Phase I of the program, the simplex package evaluation, consisted of analyses, aircraft modifications, laboratory tests, and flight tests performed by McDonnell Aircraft Co. (MCAIR) to evaluate the LTV simplex package used to power the stabilator on the F-4 aircraft.[1] The program objectives are illustrated in Figure 41.

The principal objective of Phase I was the flight test of the simplex package to determine whether the technique could be considered in the F-15 design, which at that time was undergoing development by MCAIR. This phase was planned to demonstrate the feasibility, assess the survivability, reliability, and maintainability, and form a confidence base with the F-4 aircraft. In addition, it provided experience and testing requirements for the survivable stabilator actuator package for Phase IIC.[2]

The flight test program consisted of seven flights covering most of the speed, altitude, and maneuvering range of the F-4 aircraft. Engagement and disengagement of the emergency hydraulic system (EHS) were accomplished at both subsonic and supersonic speeds. Cruises for periods up to 1 hour on EHS were conducted. Windup turns, roller-coaster maneuvers, approaches to 1 g stall, and simulated and actual landings were performed. Flight scenarios are presented in Figure 42 with testing results summarized as follows:

Control Capability: Adequate backup system performance with EHS, except in region of high hinge moments
Gear and Flap: No control problems with EHS transition
Temperature: Maximum EHS temperature 245° F.
Landing: Control with EHS adequate

The simplex package was designed to provide only a get-home-and-land capability following the loss of aircraft control hydraulic power. Flight testing indicated that this limited design objective had been exceeded, with a

---

1. The data for this discussion is taken from Report AFFDL-TR-70-135 prepared by MCAIR.

2. The SSAP for Phase IIC was a duplex design, and the first design to have a redundant EHS (emergency hydraulic system).

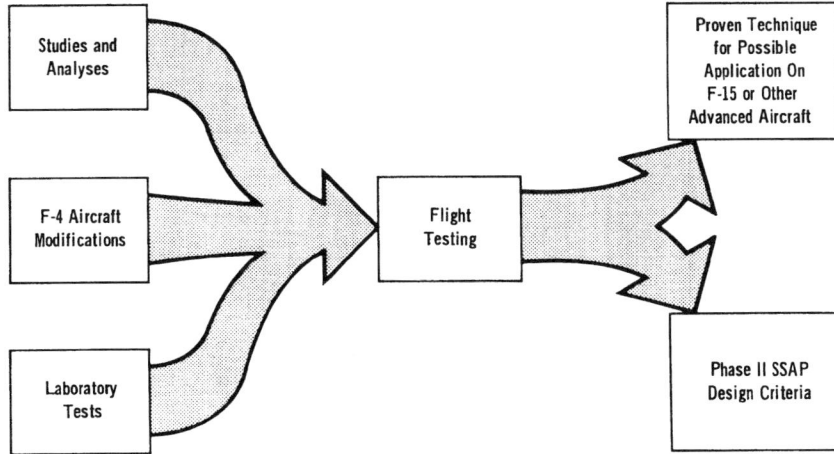

Fig. 41   SFCS Phase I program and objectives for F-4 with simplex integrated actuator package.

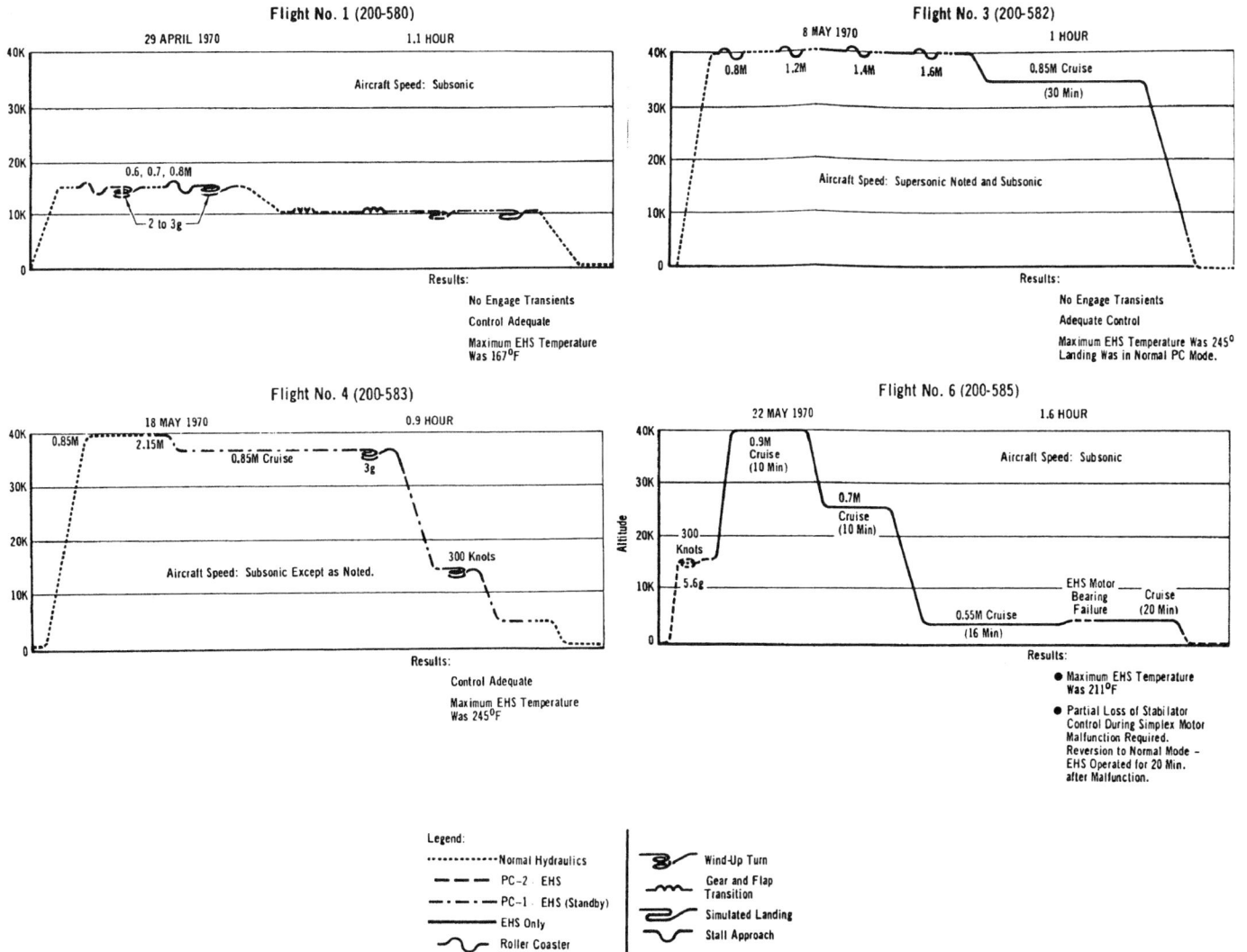

Fig. 42   SFCS Phase I flight scenarios.

substantial use envelope available with the package operating in the backup mode. However, the flow and hinge moment capabilities of the EHS preclude high acceleration maneuvering at flight conditions requiring high hinge moments or rapid stabilator motions for recovery to 1 *g*, as well as limiting the gust level under which a safer landing is assured.

A major area of concern with the simplex package test program was the possibility of high temperatures in the EHS. The use of a small volume, closed circuit emergency hydraulic system was predicted to result in fluid temperatures that would tax the capabilities of Mil-F-5606 fluid. Both ground and flight tests resulted in package temperatures below the limits of this fluid.

Survivability analyses indicated that the simplex package in the pitch axis improves the longitudinal control system survivability under the conditions examined by approximately 50% over the standard F-4 aircraft (Figure 43).

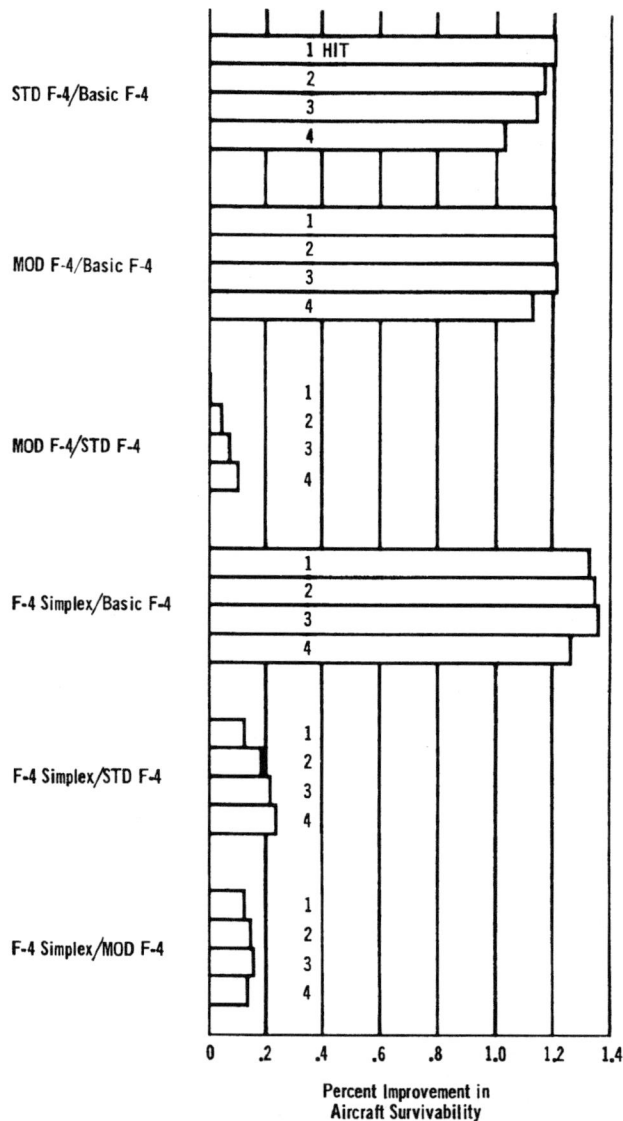

Fig. 43   Improvement in F-4 survivability.

Once the capabilities and limitations of the simplex package were evaluated, conclusions and recommendations for future package designs could be provided. The data substantiated that the simplex concept was an available technology for incorporation in the flight control systems of the aircraft being designed in 1970.

Simplex Package as a Substitute for Standard Actuator

The F-4 standard stabilator actuator is a one-piece dual tandem cylinder design of aluminum, with a slide valve that controls the flow from central hydraulic systems PC-1 and PC-2. The production F-4 stabilator actuator is capable of a maximum no-load stabilator rate of approximately 25°/second and can provide a zero-rate stabilator hinge moment of approximately 700,000 pound-inches.

The simplex was designed with features to make it more survivable than the standard actuator. It is a two-piece design with ripstop construction, which prevents a crack (as could be caused by a shell fragment) from propagating from one section to the other and causing a loss of fluid from both. In normal operation, the simplex package, like the standard actuator, receives hydraulic power from the aircraft's central hydraulic systems PC1 and PC2. In addition, the simplex has as an integral part an emergency hydraulic system that operates in the event both aircraft systems are lost, providing a get-home-and-land capability (see Figure 44 for the package's hydraulic diagram). The EHS is composed of motor, pump, reservoir, switching valve, two pressure switches, and other components.

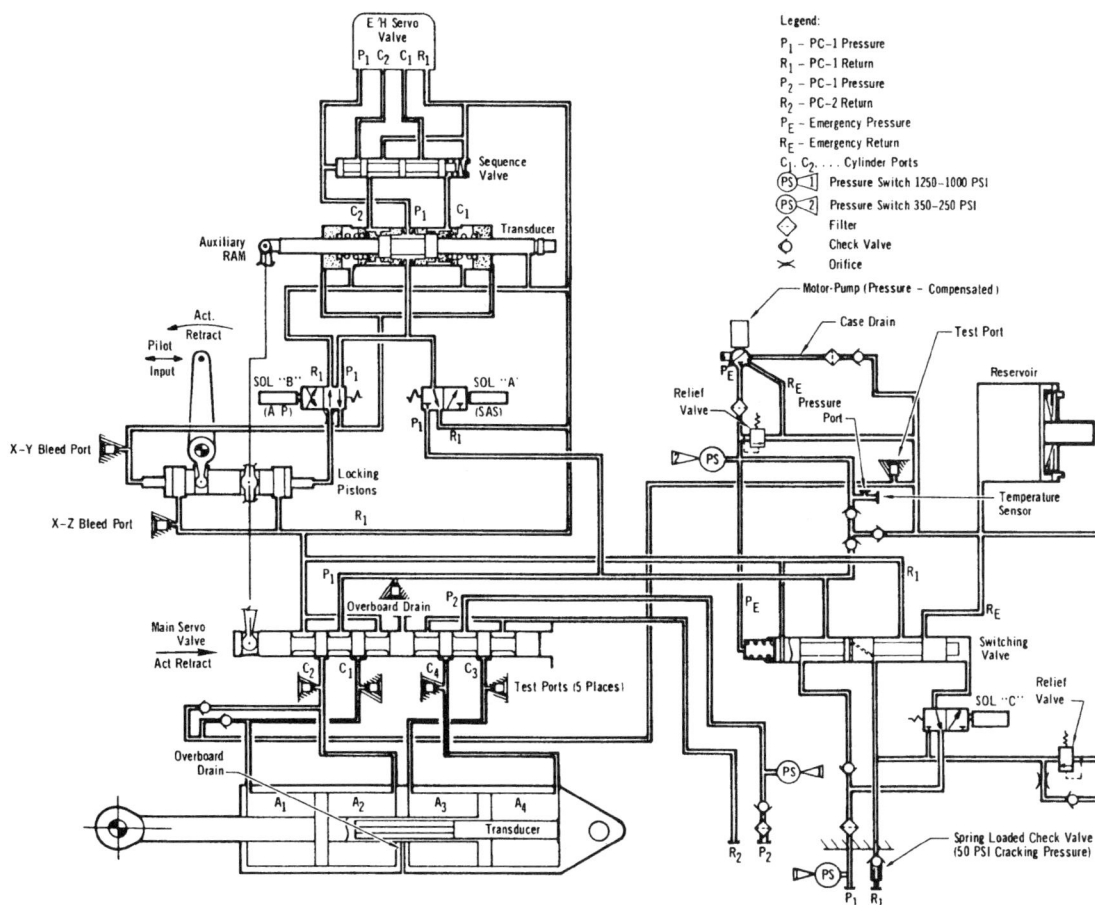

Fig. 44   Hydraulic diagram—simplex package.

Emergency Hydraulic System (EHS) Operation

In the EHS, the loss of central hydraulic system pressure is sensed by pressure switches located in the PC1 and PC2 pressure ports. During operation, when the pressure in either pressure switch drops below a safe level, the switch closes and electrical power is supplied to the motor pump. If PC2 pressure is lost, the EHS goes into standby. If PC1 pressure is lost, a switching valve ports the EHS output into the PC1 portion of the package and thence to the actuator. The package was not designed to permit porting the EHS output into PC2 or that part of the actuator.

On the flight test program, the pilot was provided the capability to simulate failure of the PC1 and/or PC2 central hydraulic systems. This was accomplished by adding a separate set of valves and hydraulic circuitry to the simplex package. A pump housing temperature of 211°F was the maximum observed in flight with EHS at altitude of 39,000 feet.

Summary of Temperature, Flutter, and Dynamic Analyses

The limitation of the F-4 flight envelope by over-temperature of the simplex package operating in the emergency mode depends upon the stabilator duty cycle demanded, the atmospheric temperature locally, the flight durations at the higher Mach numbers, the get-home-and-land flight time, and whether depletion of the hydraulic fluid is permitted.

Based on temperature analysis, the simplex package was shown to be thermally compatible with a flight envelope significantly in excess of that required solely for get-home-and-land purposes when installed in an F-4 aircraft.

An analysis was conducted to determine magnitude of changes in stabilator elastic restraint and effective stabilator pitch inertia provided by added actuator mass of a new actuator. Results from dynamic and aeroelastic trend studies concluded that the production Slotted Leading Edge stabilator powered by the simplex package possessed more than the required 15% velocity margin of safety for flutter throughout the F-4 flight envelope.

Conclusions and Recommendations of Phase I

Conclusions

1. The simplex package and the modifications to the F-4 test aircraft for installation and operation of the package did not add any single points of failure to the longitudinal flight control system.
2. The use of the simplex package increased the probability of success for a 1-hour flight.
3. The simplex package in an F-4 as compared to the Standard F-4 increased the survivability of the aircraft by approximately 0.2% and increased the survivability of the longitudinal flight control system by approximately 50%.
4. The simplex package was thermally compatible with a flight envelope significantly in excess of the subsonic get-home-and-land capability.
5. The restraint of the F-4 Slotted Leading Edge stabilator provided by the simplex package resulted in a flutter margin of greater than 15% for the F-4 aircraft.
6. No incompatibilities were found between the simplex package and the F-4 hydraulic and electrical system.
7. The simplex package operating on EHS only provided sufficient maneuvering capability to allow use of a significant portion of the F-4 flight envelope.
8. The simplex package concept was a proven technology for use in 1970s aircraft.

Recommendations

1. Design should use noncritical components to shield critical components where possible.
2. Critical components should be redundant.
3. The reservoir should have a high enough base pressure to provide adequate flow at low pump pressures.
4. A flow-through reservoir should be used to reduce case drain temperatures in hot environments while increasing reservoir temperature in cold environments.
5. Any pressure-operated lockup or switching devices should be insensitive to source-pressure transients.
6. Large packages should be grounded-body, moving-piston type to reduce effects of large swept volumes, high inertias, and susceptibility to jamming.
7. Antirotation devices should be incorporated to reduce the effects of mass unbalance.
8. Pressure-sensitive switches should be located hydraulically to avoid trapped high-pressure fluid.

Comments on Phase I

- MCAIR proved that the simplex package with its EHS provided a get-home-and-land capability and that was the main objective, so how much more could be asked?
- MCAIR didn't have duty cycle data on the F-4 longitudinal axis, but made no attempt during the flight tests to obtain the data.
- There is no recording or listing of the total operating time on EHS during flight tests.
- There is no mention of pre- or post-flight inspection or testing of the EHS.
- Mention is made of a pump bearing failure due to the grease used, but the reports fail to mention the corrective action taken and the flight test that followed.
- Apparently, there was no detailed flight plan—it appears the pilot just got in and flew.
- Any pilot would (or should) know that, with only one-fourth the hydraulic power available for control from the EHS, not all maneuvers are possible. Thus, the "unscheduled reversion required" reported for the 7 g turns on Flight No. 2 ought to have been expected.
- How do the supersonic speeds and roller-coaster maneuvers at 40,000 feet on Flight No. 3 demonstrate a get-home-and-land capability for the EHS?

Survivability Technology

The Air Force F-4E fleet was retrofitted with an emergency backup hydraulic unit that could furnish hydraulic power to the stabilator in the event power from the primary system was lost. This was a motor/pump and valving assembly located in close proximity to the stabilator that was only operated in case of an emergency.

Phase II

Upon completion of Phase I, work started on modifying the program's F-4 aircraft for installation of the equipment and changes for the Phase II flight tests. Also, at the start of this phase, a series of analyses were conducted to provide the data for the remaining program objectives, as follows:

IIA	Fly-by-Wire with Mechanical Backup
IIB	Fly-by-Wire without Mechanical Backup
IIC	Fly-by-Wire with Survivable Stabilator Actuator Package in Pitch Axis
IID	Demonstration Flights
IIE	Precision Aircraft Control Technology and Reliability/Maintainability of SFCS Fly-by-Wire

Phase IIA: Fly-by-Wire with Mechanical Backup

The early work on Phase IIA was concerned with developing the control laws for the longitudinal and lateral-directional axes of the F-4 aircraft. Some researchers questioned whether these analyses were necessary, because MCAIR had built hundreds of F-4 aircraft and knew the aircraft's aerodynamics to the *n*th degree. But such thinking is only correct when the design remains the same. Fly-by-wire changed many items on the aircraft. And the aircraft in question was not a laboratory model, but rather was the test-bed vehicle of a fly-by-wire system meant to prove the system's advantages.

The transition from a mechanical system to an electrical one can only be appreciated by making a detailed comparison on an item-for-item basis. The electronic circuitry was performing computations and providing data that had never been applied as it was here; consequently, the only proper procedure was to perform analyses and establish the gains and aerodynamic parameters that should be used.

One must also recognize that things changed considerably for the pilot. When the fully powered system came into being the pilot no longer had to apply the forces; and, because the system was irreversible, the pilot could no longer feel or tell what was going on at the control surface, necessitating artificial feel. However, with fly-by-wire the friction and inertia of the system were gone, so pilots had to learn all over again, now using a side-arm controller in place of the center stick. Figure 45 shows the location of the Survivable Flight Control System components in the F-4 aircraft.

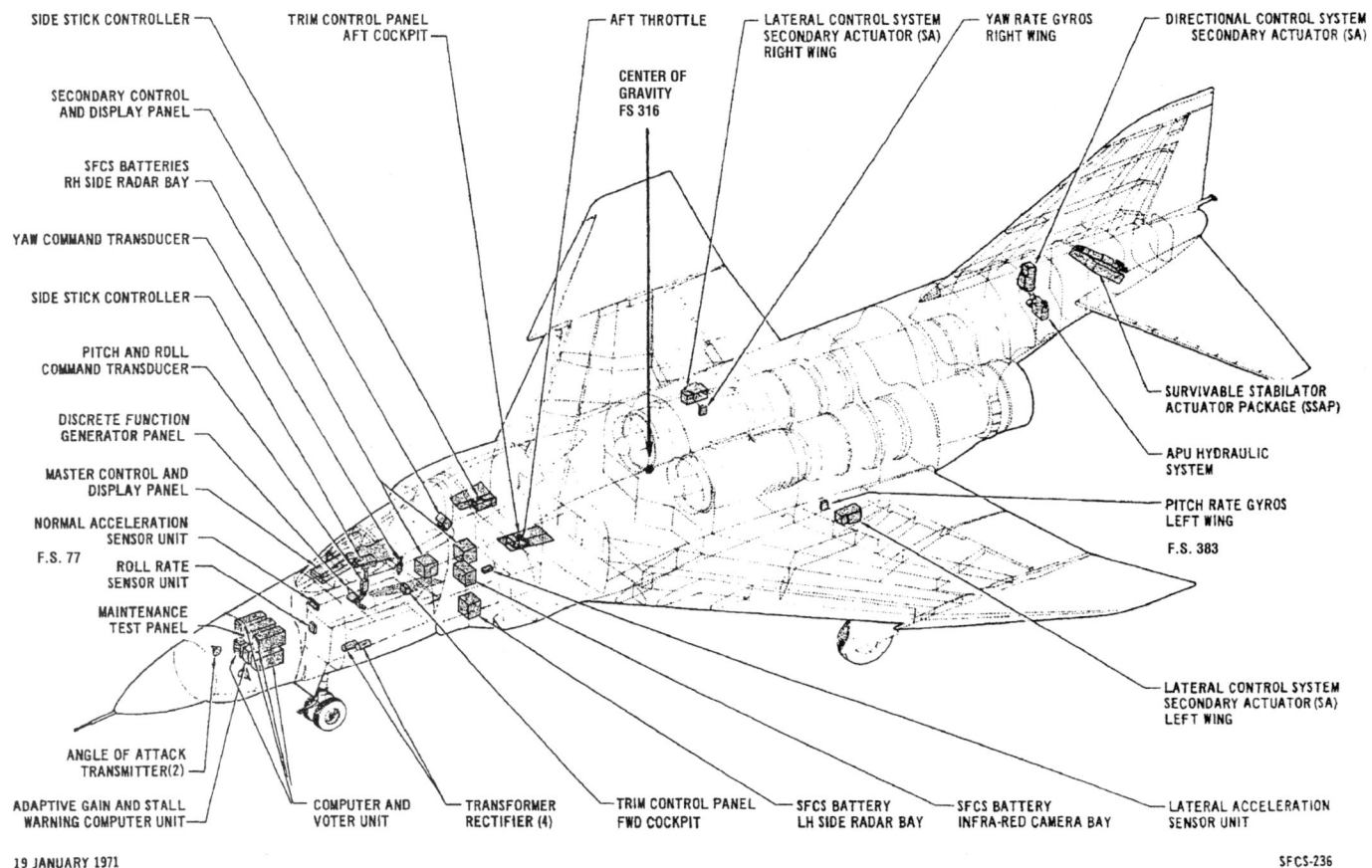

Fig. 45   SFCS equipment location.

Mechanization of Actuation

Figures 46 and 47 must be used if one is to have a clear understanding of the components and how they operated in Phases IIA and IIB. The F-4E aircraft normally has three engine-driven hydraulic power supplies—PC1, PC2, and utility—and an electric-motor–driven auxiliary power unit (APU) that provides hydraulic power to the stabilator actuator in the event of an emergency. The secondary actuators on Phase II required four independent hydraulic supplies, so the APU was converted to full-time use as the fourth hydraulic power source.

As shown on Fig. 46 the lateral axis used two secondary actuators to control the ailerons and spoilers; one actuator controlled those on the left-hand side and the other those on the right-hand side. These actuators received their input from the fly-by-wire system and their outputs drove the aileron and spoiler actuator control valves. The stabilator and rudder operated through a mechanical device called the mechanical isolation mechanism. It was pilot-operated: in one position, the pilot had full control to the input linkage on the stabilator actuator; in the opposite position, the pilot had no control over this linkage and the input to the stabilator actuator came from the secondary actuator, which in turn received its input from the fly-by-wire system. Because the operation of this arrangement was left strictly to the pilot's discretion, the two positions were called the manual and fly-by-wire modes.

On Phase IIA then, the lateral axis was controlled by fly-by-wire secondary actuators and the longitudinal and directional axis through the mechanization as described. In Fig. 47 the mechanical isolation mechanism has been removed and the secondary actuators control movement of their respective slide valves by fly-by-wire inputs to the actuator.

Hydraulic Fluids for Survivable Flight Control Systems

MCAIR's analysis of the various hydraulic fluids that could be used on the SFCS vehicle included a trip to the Air Force Materials Laboratory to obtain the current data on the most recently developed fluids. Based on their analysis, the Mil-H-83282 fluid was selected:

1. Mil-H-83282 has a low viscosity at −40°F and consequently makes cold starts easier. Mil-H-83282 also has the higher viscosity at 450 °F, which gives lower leakage rates and better lubrication.
2. The combustion indices and flammability test results showed Mil-H-83282 is significantly less flammable.
3. The results of fourball wear tests indicated that Mil-H-83282 causes less wear at the higher loads.

All flight tests on the F-4 aircraft in Phase II used Mil-H-83282. This was the first real test of this fluid in a series of laboratory, ground, and flight tests. Test results showed this to be a superior fluid, especially in the higher temperature regime. The SFCS program became the initial benchmark for Mil-H-83282, and by 25 years later it had become the standard fluid used by the Air Force in military hydraulic systems. When used in flight control systems it shows a higher bulk modulus at higher temperature—a very good attribute because it thus produces a higher actuator spring rate and in turn reduces the problem of control surface flutter.

After the Air Force adopted Mil-H-83232 as the standard, Mil-F-5606, the old red oil,[3] so long familiar to Air Force maintenance personnel, could finally vanish. When a hydraulic system fluid was replaced with Mil-H-83282, the

---

3. The Mil-F-5606 hydraulic fluid had red dye so that leaks could be more easily located.

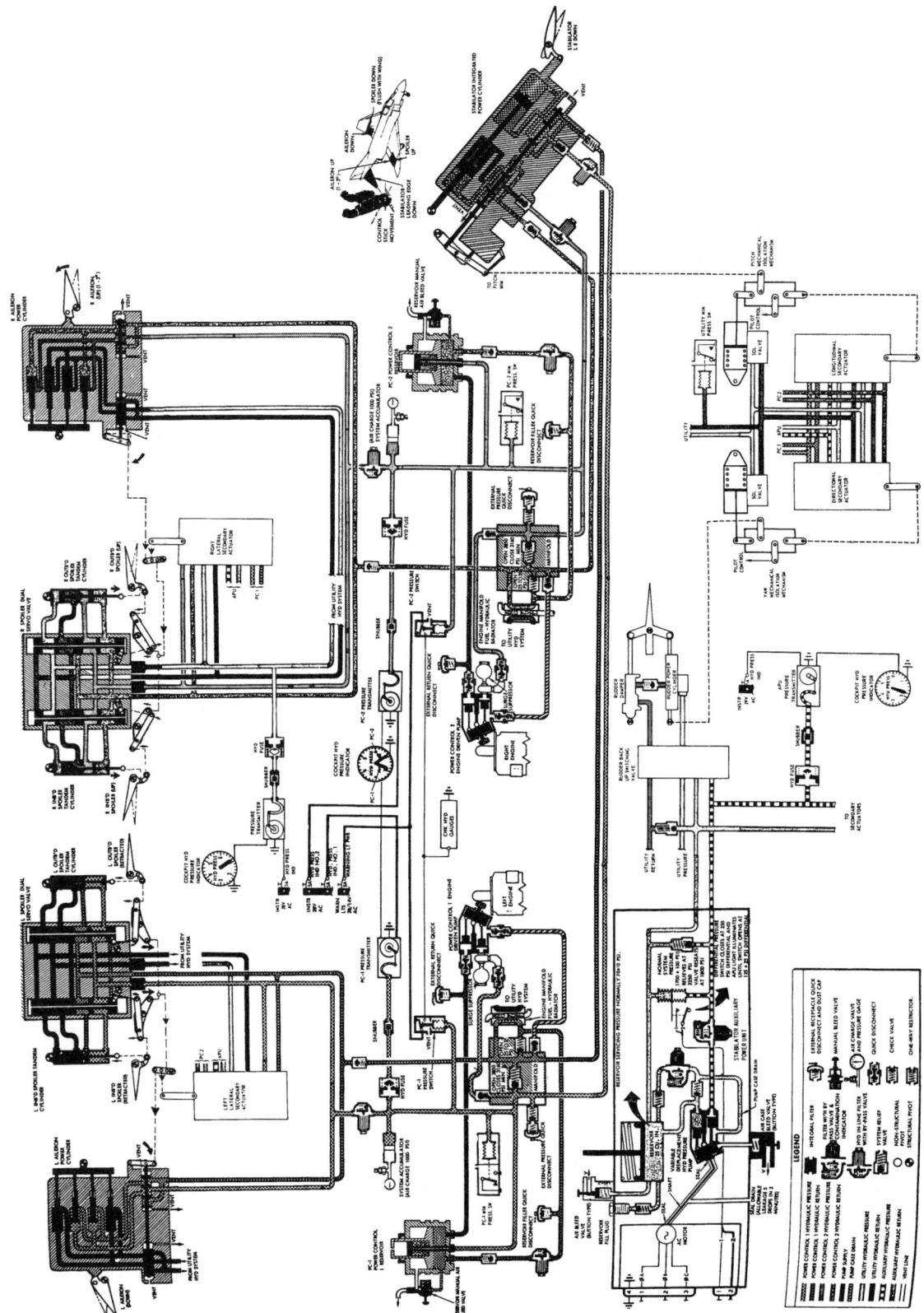

Fig. 46   SFCS flight control hydraulic system phase IIA.

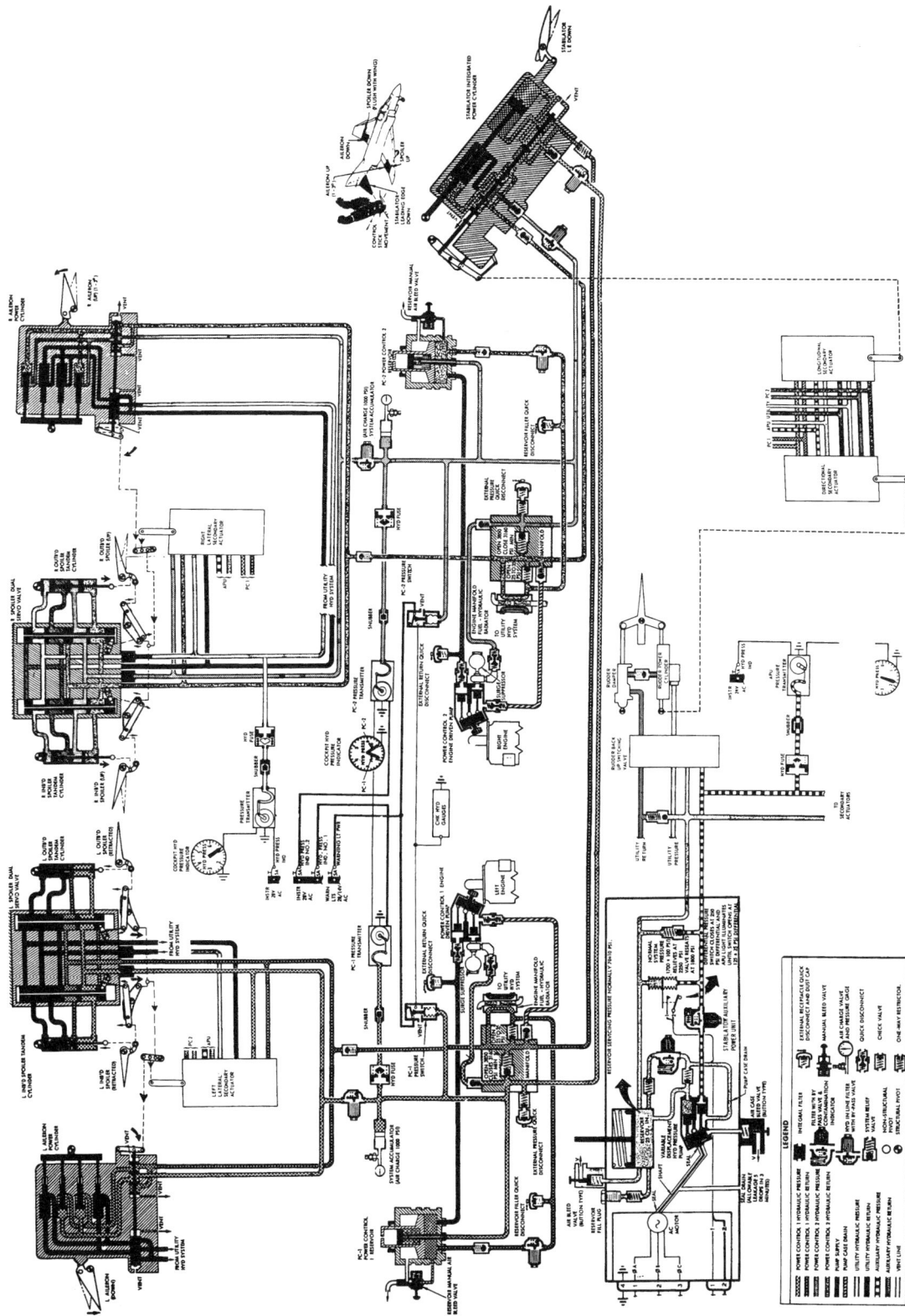

Fig. 47   SFCS flight control hydraulic system phase IIB.

residual Mil-F-5606 seldom created any problem as the two fluids are quite compatible. By using Mil-H-83282 and proving its superiority over Mil-F-5606 by flight tests, the SFCS program provided the Air Force with a benefit that is seldom, if ever, mentioned in any discussions of fly-by-wire.

SFCS Handling Quality Analysis

Fly-by-wire aircraft are often regarded as highly mechanized and loaded with electronic gear and controls. Yet, by definition, fly-by-wire is a flight control system designed so that aircraft *motion* is the controlled variable. Unless we are talking about drone aircraft, a human pilot must be in the control loop. The pilot is the source of commands, and the aircraft's response to these commands gives rise to the pilot's feel system. Consequently, in any fly-by-wire system, the handling qualities afforded the pilot have a vital role. MCAIR performed an analysis of the handling qualities on the Survivable Flight Control System.

Handling Quality Data

Past programs for development of longitudinal and lateral-directional handling qualities were directed toward establishing limiting values of the traditional performance parameters (such as frequency, damping, and time constants) that pilots feel are consistent with desired levels of precision and control during maneuvering flight. Past work was used to update military specifications, and was directed toward specifying handling qualities for aircraft that do not use aircraft motion feedbacks in the primary flight control mode. The introduction of highly augmented flight control systems and fly-by-wire systems such as the SFCS questioned the adequacy of existing specifications and performance criteria. An investigation was conducted to define short-period performance criteria requirements.

The longitudinal and lateral-directional controls were implemented to comply with the requirements of Mil-F-8785B (ASG) and the C* criterion. A handling qualities criteria development effort was conducted concurrently with the control law development program.

A six-degree-of-freedom man-in-the-loop simulation program using a fixed-base flight simulator was conducted to evaluate the control law implementation. This simulation included the capability to maneuver the aircraft throughout the F-4E flight envelope and stall and post-stall conditions.

The SFCS handling quality requirements were defined as follows:

- To what degree are C* handling qualities criteria compatible with the required mission loop closures?
- How do higher order and nonlinear characteristics affect application of C* criteria?
- Can lateral-directional handling and flying qualities be incorporated into a new criterion?
- Should control laws be based on mission modes or tasks rather than the traditional short-period handling qualities and control techniques?
- Is inter-axis coupling desirable? To what degree?

- **Response Characteristics**   The response characteristics are C* and dc^2/dt criteria. The time history responses of seven aircraft conditions are given in Figures 48 and 49. Flight conditions represent extreme points on the flight envelope included in the analysis primarily to investigate achievable stability margins.

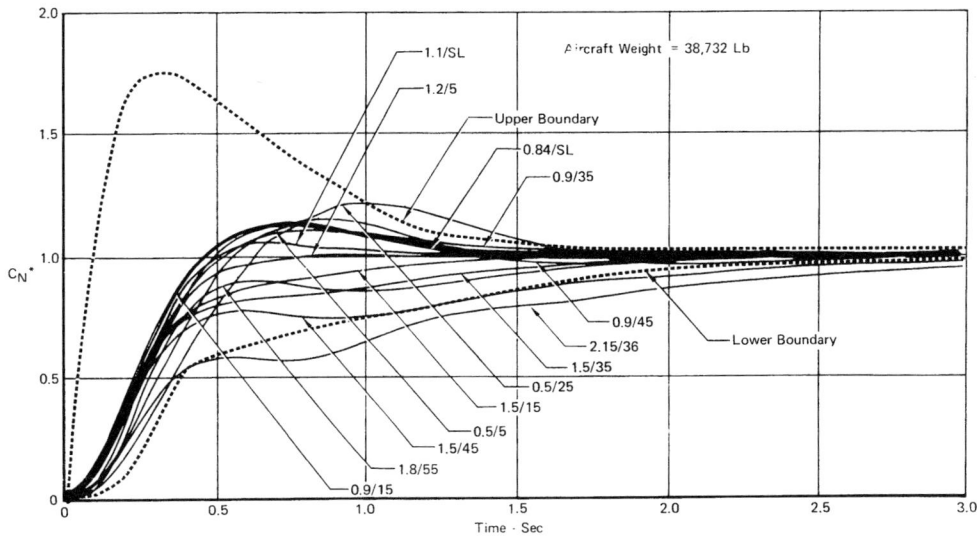

Fig. 48 Longitudinal SFCS time history C* criteria compliance.

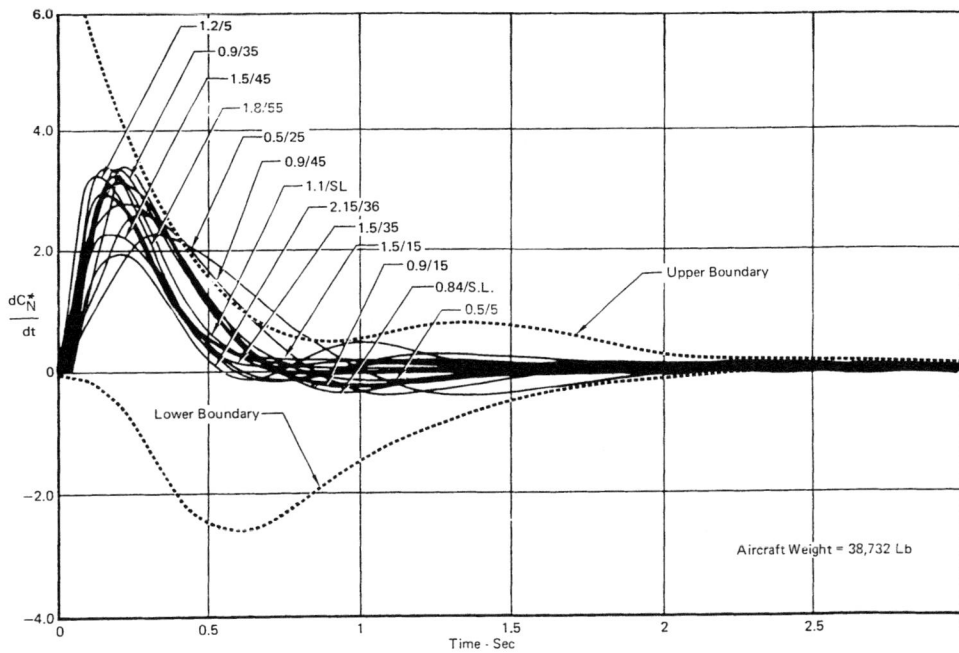

Fig. 49 Longitudinal SFCS time history $\frac{dC*}{dt}$ criteria compliance.

The handling qualities of the normal neutral speed stability (NSS) mode were found to be quite satisfactory by all pilots evaluating the SFCS simulator, and received an average Cooper rating of 2.

An expression designated D*, which is related to side-slip angle β, was derived and analyzed to determine any potential application to handling qualities.

SFCS Side-Stick Controller

A mock-up of side-stick controller used on the SFCS program was evaluated during certain flight tests on the simulator. It was installed on the right console of the crew station cockpit and was used to perform various tasks. It was found to be an acceptable flight controller. Pilots felt the initial breakout forces were too high, and they were reduced to a desirable level. The simulator results indicated that flight tests with real side-stick hardware were needed before valid comparisons could be made between controllers. A time history of a landing approach using the side-stick controller in the normal mode is shown on Figure 50. Cumulative distribution plots indicating a slight variation in performance with the center and side-stick controllers is shown in Figure 51.[4] This plot shows that the side stick is an effective flight controller for landing approach. Pilots used the side stick for other tasks and found it was usable in all cases.

Just as in the case of hydraulic fluids, the SFCS program provided an invaluable benefit to advancing the use of side-stick controllers. The very same side-stick controller used for SFCS was adopted for use in the YF-16—same supplier, same design (only the springs were changed so that the YF-16 stick seemed to have virtually no motion). This is yet another significant payoff of the SFCS program that most discussions overlook.

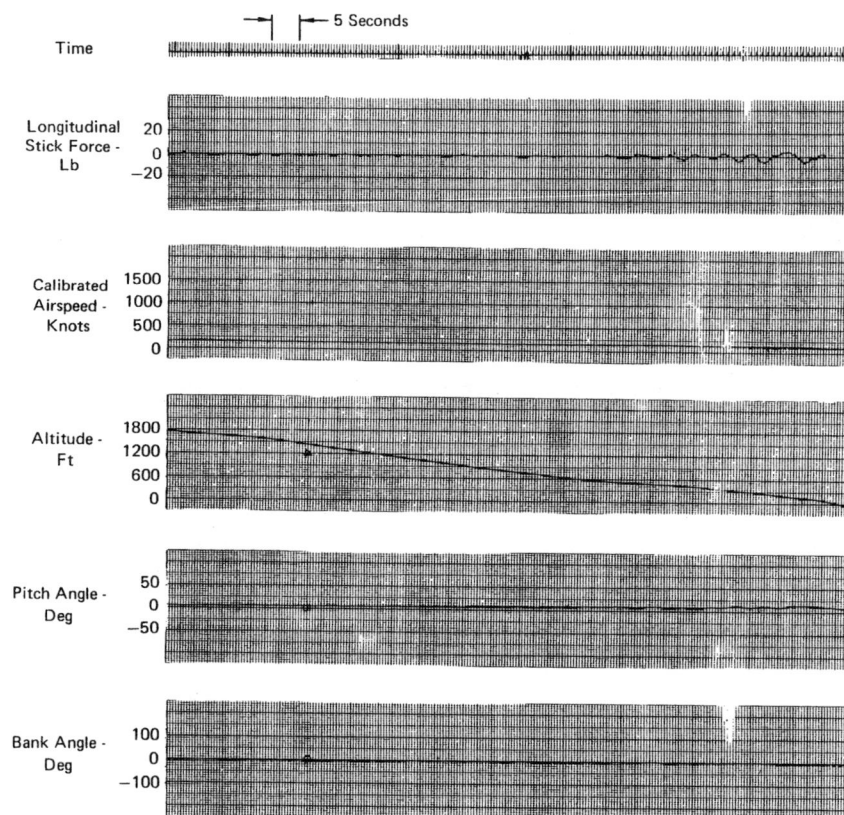

Fig. 50   Landing approach using side-stick controller.

---

4. Fig. 51 is one of the few times comparison data has ever been presented, either by simulator or actual flight test.

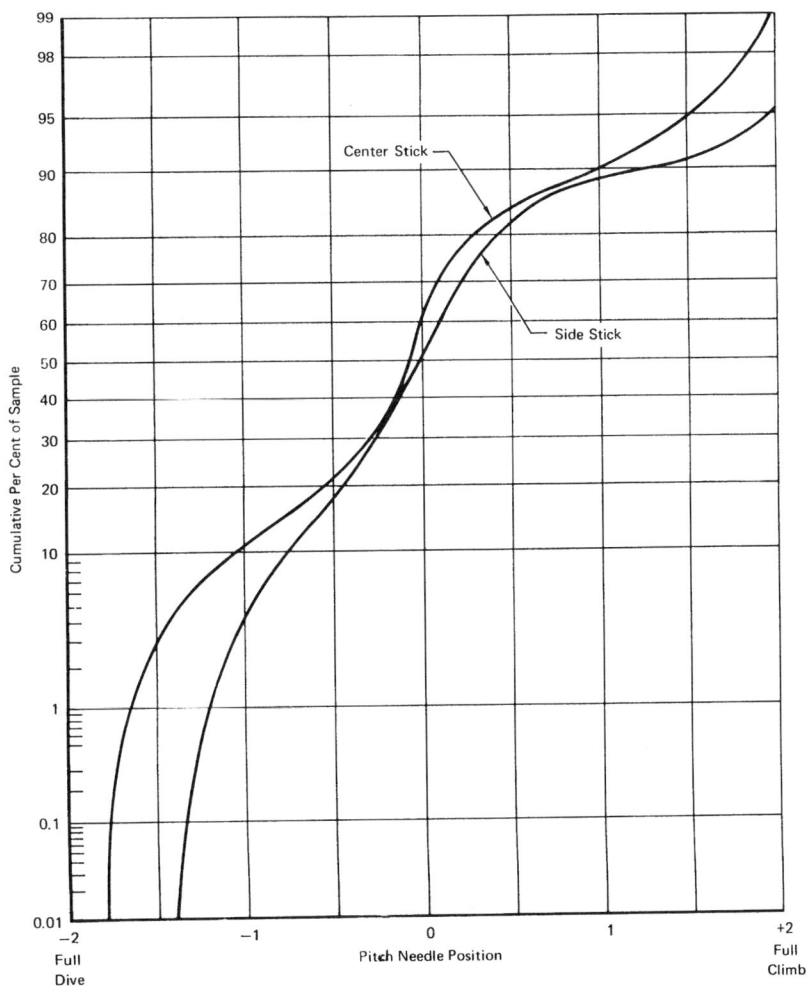

Fig. 51   Terrain following task cumulative distribution for SFCS using center stick versus side-stick controller.

Pilot Comments on Phase IIA

The most useful comments that can be made on handling qualities in relation to fly-by-wire can be found in Appendix B, including commentary by the MCAIR flight test pilot on the F-4 aircraft #12200 who made over 20 flights that covered the entire flight envelope on fly-by-wire.

SFCS Secondary Actuators

There are two important points on the electrical signal transmission portion of a fly-by-wire system—where the signal starts and where it ends. On the SFCS system it is multichannel, quadruple signals generated by electrical transducers mounted on the pilot's control stick. The four channels transmit independently of each other, and physically end at an actuator having four channels or independent sections. The output of these four channels is summed physically at a single output shaft. The actuator that performs this conversion is called a *secondary actuator,* because it drives the control valve on the main or primary actuator that drives the flight control surface. The secondary actuator is a quadruplex, force-summing, electrohydraulic servomechanism. It is a self-contained unit consisting of four independent servo-controlled elements coupled to a common output. Figure 52 shows one mechanized element.

Notes:
1. Element output data (nominal):
   Piston area · 0.294 sq in.
   Stroke · ± 0.500 in.
   Force · ± 294 lb
2. Differential pressure sensor
   detection levels (nominal):
   90 psi · loss of pressure
   714 psi · initiation of motion
   930 psi · failure level
   1000 psi · relieve pressure to return
3. The centering/braking release piston
   is shown in the locked/braked position.
   The centering/braking mechanism is
   spring driven and pressure released.
4. In case of element failure, pressure is
   shut off by the solenoid and the
   piston is by-passed through the jet
   pipe servo valve receiver.

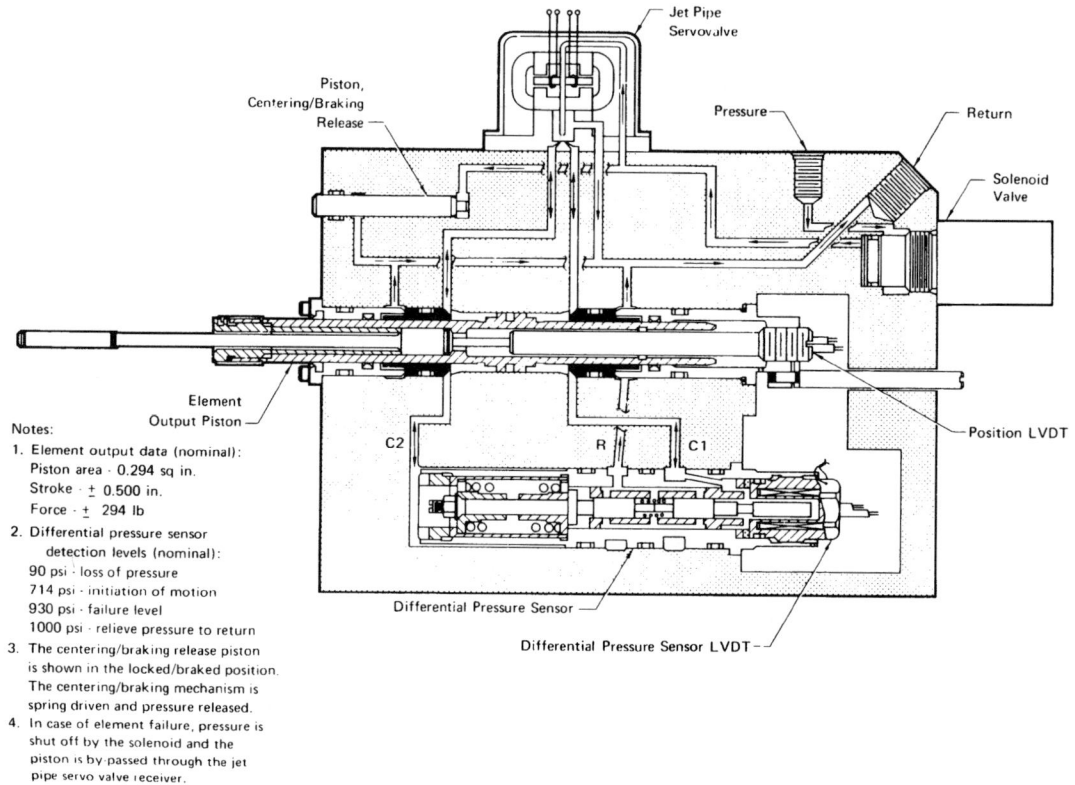

Fig. 52   Hydraulic schematic, single actuator element.

Table 1 illustrates the type and number of parameters involved in the secondary actuator performance. Analyses and laboratory tests were conducted on all the secondary actuators to confirm that they met the design, installation, and performance requirements for use on the SFCS.

Secondary actuator failure data is worthy of mention. A failure transient will result whenever an active failure occurs in an element. When such a failure occurs, the actuator output is displaced until a force balance is achieved. The force balance is maintained until the failed element is switched off-line. On third failure, a brake is applied to hold position when the last two elements are switched off-line. Secondary actuator design parameters are such that a failed element cannot overpower a good element or third failure.

Simulation

Flight simulators provide advantages for making handling assessments. Extra elements in the flight situation can be included in the assessment. Most importantly, the pilot is able to experience the proposed stability and control characteristics and possibly try out variations on these at an early stage in a new design.

A comprehensive test plan was undertaken of the pilot's performance in SFCS man-in-the-loop simulations using the results of previous analyses as a guide. A very complete hybrid mechanization was necessary to achieve high visual fidelity and maximum crew-station realism with the fixed-base equipment. This requirement was satisfied by having a major portion of the software and flight control system data programmed on a digital computer that interfaced through an analog computer to the crew-station hardware. Each pilot was required to fly a baseline mission with various configurations and then provide a Cooper-Harper rating for each of five scored mission phases.

- The SFCS flight simulator used a fixed-base crew station equipped with both center-stick and side-stick controllers and instruments for three-axis aircraft maneuvering throughout the flight envelope.

68

Table I
Secondary Actuator Parameters

Symbol		Value	Units
A_V	Effective secondary actuator piston area	0.294	In^2
B_4	Linkage freeplay secondary actuator elements to summing link	0.002	In
B_5	T-valve hysteresis	0.0	Ma
B_6	Hysteresis due to friction effects	0.047	Ma
C_D	Ratio of flow gain to pressure gain	0.000369	Cis/Psi
F_F	Coulomb friction (for 4 elements)	16.0	Lb
F_1	Force output of element 1 of secondary actuator	*	Lb
F_2	Force output of element 2 of secondary actuator	*	Lb
F_3	Force output of element 3 of secondary actuator	*	Lb
F_4	Force output of element 4 of secondary actuator	*	Lb
Gi	Preamplifier gain	0.0142	Ma/V
Gf	Filter gain	0.05	Ma/V
H	Feedback gain, $K_X K_{dm} K_F$	5.19	V/In
K_a	Servoamplifier gain	57.6	Ma/V
K_{dm}	Demodulator gain	1.25	VDC/VAC
K_F	Feedback amplifier gain	0.296	V/V
K_L	Open loop gain	122	Sec^{-1}
K_{p1}	Pressure gain for 3000 psi supply	325	Psi/Ma
K_{p2}	Pressure gain for 1400 psi supply	152	Psi/Ma
K_S	Structural spring constant	1.46×10^5	Lb/In
K_V	Servovalve gain	0.12	Cis/Ma
K_X	LVDT scale factor	14.0	V/In
L	Total secondary actuator piston stroke	1.0	In
L_4	Linkage ratio—summing link to elements of secondary actuator	1.37	In/In
M	Effective mass of linkage	0.456	$Lb\text{-}Sec^2/In$
N_1	Number of operating elements on 3000 psi	*	*
N_2	Number of operating elements on 3000 psi opposing a failed element or an element with a large tolerance buildup	*	*
$P\Delta P$	Pressure differential across a secondary actuator piston	*	Psi
P_M	Maximum differential pressure across a secondary actuator piston	1000	Psi
P_R	Return pressure	*	Psi
P_S	Supply pressure	*	Psi
P_T	Tripout pressure of differential pressure sensor	930 Psi	Psi
Q	Flow	*	Cis
Q_{SA}	Flow rate for secondary actuator element	*	Cis
V_e	Error signal into servoamplifier	*	V
V_i	Input command signal	*	V
V_S	Signal in one element due to tolerance buildup	*	V
X_1	Displacement—secondary actuator elements	*	In
X_2	Displacement—secondary actuator summing link	*	In
ξ	Damping ratio	*	*
τ_F	Feedback amplifier time constant	0.001	Sec
τ_V	Servovalve time constant	0.00177	Sec

*The value of the parameter is specified elsewhere in a drawing or applicable specification.

- Also included in the flight simulator were the instruments necessary for control of aircraft altitude, angular rate, and velocity during flight.
- Solution of the equations-of-motion and coordinate transformations generated signals to drive the following instruments:

  Altitude/direction indicator
  Airspeed/Mach meter
  Angle-of-attack indicator
  Barometric altimeter
  Rate-of-climb indicator
  Normal load factor
  Engine tachometers
- Flight controls installed in the simulator cockpit included:

  Rudder pedals
  Throttle quadrant
  Speed brake
  Center-stick controller
- An artificial feel system was installed to provide realistic longitudinal axis stick forces on the center stick to the pilot.
- A visual display system was also used in the simulation and included a large-scale terrain map shown in Figure 53.

Fig. 53   Visual display terrain map.

Flight Test Results for Phase IIA

The first flight of Phase IIA, using YF-4E S/N 62-12200, was conducted in St. Louis, Missouri, on April 29, 1972. The final flight of this phase was performed at Edwards Air Force Base in California on September 8, 1972. A total of 27 flights were conducted, including two ferry flights as the test program moved to Edwards AFB prior to low altitude supersonic envelope expansion. The total system flying time was 30.6 hours, with 23.0 of these hours in the full three-axis fly-by-wire configuration. Of the 27 Phase IIA flights, 9 were completely fly-by-wire flights from engine start to shutdown. Acceptable performance for fly-by-wire operation was developed and adequate confidence in system reliability was established, warranting removal of the pitch and yaw mechanical backup for continued testing.

Phase IIB: Fly-by-Wire without Mechanical Backup

The first flight of Phase IIB was conducted at Edwards AFB on January 22, 1973. A total of 19 flights were flown for an accumulated 18.3 hours. The performance of the fly-by-wire flight control system with the mechanical backup removed was cross-checked throughout the Phase IIB flight envelope. Emphasis was placed on the evaluation of lateral-directional handling quality characteristics. The final Phase IIB flight of May 7, 1973, was a flight envelope clearance with external fuel tanks and B.L. 81.5 pylons for a subsequent Air Force Phase IID evaluation flight. Phase IID flights were accomplished concurrently with the latter portion of Phase IIB.

Results

The SFCS aircraft was flown using fly-by-wire flight control throughout the level flight envelope of the F-4. A maximum load factor of 5.0 *g* was attained. The following maneuvers and tasks were demonstrated in addition to classical stability data collection:

- Fly-by-wire take-offs and landings, formation flying, instrument flying and approaches, "over the top" maneuvers (loops, etc.), and general flying maneuvers.
- Simulated combat maneuvers such as moderate angle-of-attack roll maneuvers, transonic decelerating turns, and air-to-air and air-to-ground tracking runs.
- In-flight reversion from fly-by-wire "normal" to fly-by-wire electrical backup (EBU) and take-off and landing in EBU.

Significant accomplishments of the Phase II flight tests include:

- First fighter aircraft flight with total fly-by-wire.
- First Mach 2 fly-by-wire flight.
- Established pilot acceptance of fly-by-wire.
- Demonstrated superior handling qualities for transonic deceleration and control of load factor during high acceleration turns and touch-and-go landings.
- Demonstrated fly-by-wire control for general maneuvering and precision flying.
- Demonstrated fly-by-wire system reliability and maintainability.
- Developed and verified feasibility of the SFCS fly-by-wire system.

A total of 49 flights were flown to complete the Phase IIA and IIB testing. The total also includes four flights for ferrying the aircraft to Edwards Air Force Base and the return to St. Louis upon completion of the program.

Flight Preparation Ground Tests

Pilot checklist procedures for preflight and flight operations were verified during a comprehensive functional check of normal aircraft systems and the SFCS. Two high-speed taxi tests were conducted prior to first flight with pitch and yaw in mechanical backup (MBU) to ascertain that no uncommanded SFCS lateral control inputs were generated by runway take-off roll environment. A third taxi test verified safe fly-by-wire control in pitch, roll, and yaw simultaneously.

1. *Pitch actuator loop stability tests* were conducted to verify dynamic stability characteristics of the surface control loop including the pitch SA (secondary actuator), pitch MIM (mechanical isolation mechanism), and stabilator actuator.
2. *Control system proof tests* were accomplished to ascertain the strength capability of the modified control system linkages and associated support structure modifications.
3. An *electromagnetic compatibility test* was conducted with the SFCS as both source and victim. No electromagnetic interference of significant magnitude was identified.
4. *Closed-loop performance tests* were conducted using the MGTF (mobile ground test facility) with the SFCES (survivable flight control electronic system) in the aircraft. Static tests verified loop gains, sensor output gradients and phasing, control surface authority, and various discrete functions. Dynamic tests checked the flight control system closed-loop response, gain margin, and limit cycle characteristics.
5. *Control system loads measurement*—control system loads due to BIT (built-in test) operation and preflight manipulation of the cockpit controllers were measured to verify that design strength could not be exceeded.
6. *Functional ground tests* with aircraft engines running were accomplished in conjunction with taxi tests prior to the first flight

Phase IIA Flight Tests

Flight test development and evaluation of the SFCS was accomplished as a progressive buildup verification of the various operational modes and functions. Tests of the MBU mode were limited to only those required to ensure aircraft controllability for reversion to that mode. Initial test flights consisted of a systematic functional check of the SFCS and basic aircraft systems within a limited flight envelope. Tests were then oriented toward evaluating specific handling qualities in all axes while expanding the flight envelope to normal F-4 limits.

Phase IIB Flight Tests

The operation of the FBW flight control system was evaluated throughout the Phase IIB envelope. This envelope was reduced slightly from normal F-4 limits as agreed during Air Force safety reviews prior to the start of testing. The flight enveloped and the IIB points evaluated are documented in Figure 54. Maximum aim values of 20 units AOA (angle of attack) and 5 g normal acceleration were also established during the Air Force reviews. These limitations were imposed to provide an overall conservative approach to the testing and are not traceable to aircraft or SFCS configuration characteristics.

Survivable Flight Control System Test Envelope

Longitudinal flying qualities were rechecked at selected flight conditions; however, emphasis was on the quantitative assessment of lateral-directional handling qualities. Other activities included the evaluations of maneuvering and precision flight tasks. Air-to-air and air-to-ground tracking tasks were flown. Additional maneuvers included formation flight, Immelmann turns, loops, and 1/2 Cuban eights.

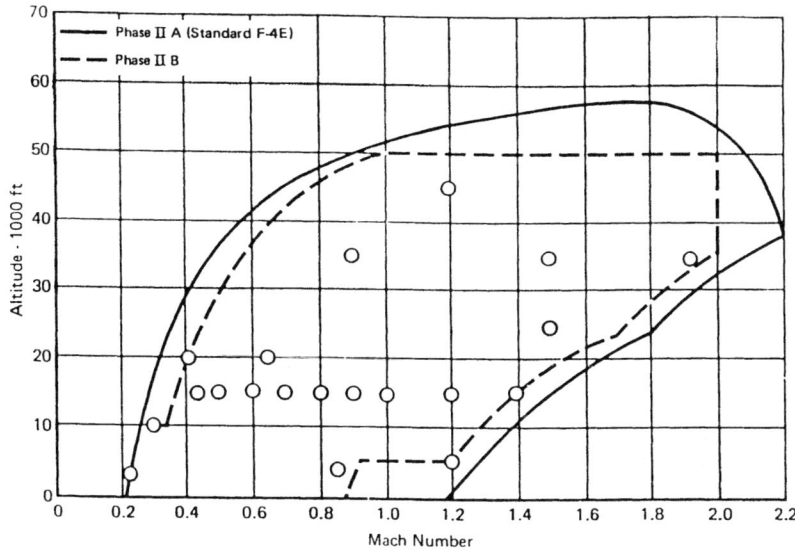

Fig. 54   Survivable flight control system test envelope.

The last Phase IIB flight was for evaluation of the SFCS performance with external tanks and pylons installed. A series of roll, pitch, and yaw maneuvers was flown, clearing the flight envelope required for a subsequent Phase IID evaluation flight.

SFCS Aircraft Flight Characteristics

Longitudinal Stability and Control

When operating in the normal fly-by-wire mode, pitch control was generally improved over the basic F-4. The pitch axis is better damped than the F-4, yet the aircraft still has adequate short-period response. Pitch short-period damping ranges from dead-beat to slightly over-damped throughout the flight envelope. The SFCS reduces the tendency to couple with the short-period motion. Stick centering is greatly improved compared to the F-4 resulting in better PA (power approach) configuration speed stability stick force cues.

The neutral speed stability (NSS) function enhanced the longitudinal control characteristics by providing automatic pitch trim to maintain 1 *g* flight through the flight envelope with landing gear up. NSS tends to reduce the pilot workload during maneuvers involving rapid airspeed or altitude changes because manual trimming is not required. Consequently, pitch control is improved as only the constant maneuvering forces are required. Effectiveness of the NSS was very obvious during the decelerating wind-up turn maneuver through the transonic area. The normal F-4 nose rise was not present and manual trimming was not required.

The center-stick maneuvering force gradients for the longitudinal SFCS were compared to the basic F-4 for several flight conditions. The data substantiates pilot comments of improved maneuvering pitch control over the F-4. The SFCS provides a more comfortable stick force gradient and stick displacement throughout the flight envelope, allowing more precise control of pitch rate and acceleration. The improved pitch control at the high acceleration values is also attributed to the overall linearity of the Fs/g gradient versus g. The SFCS stick force per g characteristics obtained from flight test data compare favorably with the predicted values determined from earlier studies and analyses.

Pitch MED gain was determined to be optimum for take-off and landing, with manual gain selected in the fly-by-wire normal mode. LOW gain was then selected at 275 to 300 knots after take-off. Take-off in fly-by-wire requires only a small aft stick force to obtain the stabilator position for rotation at lift-off. The application of aft stick forces greater than required for full stabilator can delay subsequent nose-down stabilator response. Take-off control in normal mode is good. Pitch control is excellent for landing in normal mode. Touch-and-go landings exhibited superior handling qualities and the presence of ground effect was not detectable.

The aircraft with external wing tanks installed was also well damped and exhibited no noticeable change to in-flight characteristics as compared to the clean SFCS aircraft.

Additional differences were noted in longitudinal response characteristics between the SFCS and a production F-4 aircraft. The SFCS aircraft has a tendency to maintain 1 g during stall approaches, necessitating pilot action to push the nose down for recovery. The SFCS also attempts to hold zero pitch rate at the top of a loop and the nose must be pulled down to complete the maneuver. Pilots adapted readily to these differences and they were not considered deficiencies.

SFCS System Operation

Built-In Test (BIT) and In-Flight Monitoring (IFM)

The built-in test (BIT) system is a highly complex, close tolerance system that provides a complete end-to-end check of the SFCS. During the initial flight test activity, the BIT tolerances tended to reject hardware that was operationally functional but not quite up to the performance levels expected. Engineering judgment was applied for the decision to proceed with test flights, even though BIT runs with the engines running resulted in a NO-GO. This approach did not compromise flight safety due to the extensive preflight functional checks, in-flight monitoring (IFM), and the availability of the MBU in the event of critical SFCS faults. By SFCS Flight No. 8, chronic BIT fault indications were corrected, and provision for visual observation of the BIT test sequence number was added so that decisions to proceed with an indicated NO-GO could be supported by positive assessment of the particular indicated fault. By this time, a number of replacements of SFCS equipment and/or minor changes in performance had been made. The probability of a BIT GO status improved considerably for the rest of the test program; a BIT GO with engines running was obtained prior to 17 of the remaining 20 Phase IIA flights.

Side-Stick Controller

The following discussion is an assessment of the side-stick controller based on pilot comments from the Phase IIA and IIB flight test programs.

During the SFCS flight testing all maneuvers performed using the center stick were also performed using the side stick. In general, all of the pilots found the side-stick controller to be very flyable. In fact, one of the Air Force evaluation pilots said that he preferred the deficiencies in the side-stick. The side-stick control had some characteristics that were somewhat annoying to many of the pilots, but most of these characteristics were exhibited only in specific maneuvers that required very precise control.

In evaluating the side-stick controller all pilots agreed that there was a definite learning curve associated with flying using the side stick. One of the Air Force evaluation pilots emphasized the point very well when he reported that in an aircraft with a center stick he felt a pilot can make a good evaluation in two or three flights, but in an aircraft with a side stick he felt the pilot would need to do a lot of flying before he could make a good evaluation. Because of this learning curve, the assessment of the side-stick controller was based mainly on comments by the MCAIR pilots and those Air Force pilots having three or more flights in the airplane.

Phase IIC: Fly-by-Wire with Survivable Stabilator Actuator Package (SSAP) in Pitch Axis

Current with Phases IIA and IIB was a contract to LTV E-Systems to provide an IAP for Phase IIC. Figures 55 and 56 show the components that made up the SSAP. Of the two hydraulic power supply units used in this design, only one is shown in Fig. 55. In many respects the SSAP is like having two simplex IAPs assembled to function and operate as one IAP.

MCAIR approached Phase IIC with extreme caution, as loss of their one airplane meant the end of the entire program. Caution gave rise to doubt, and doubt could only be removed by more facts and more data.

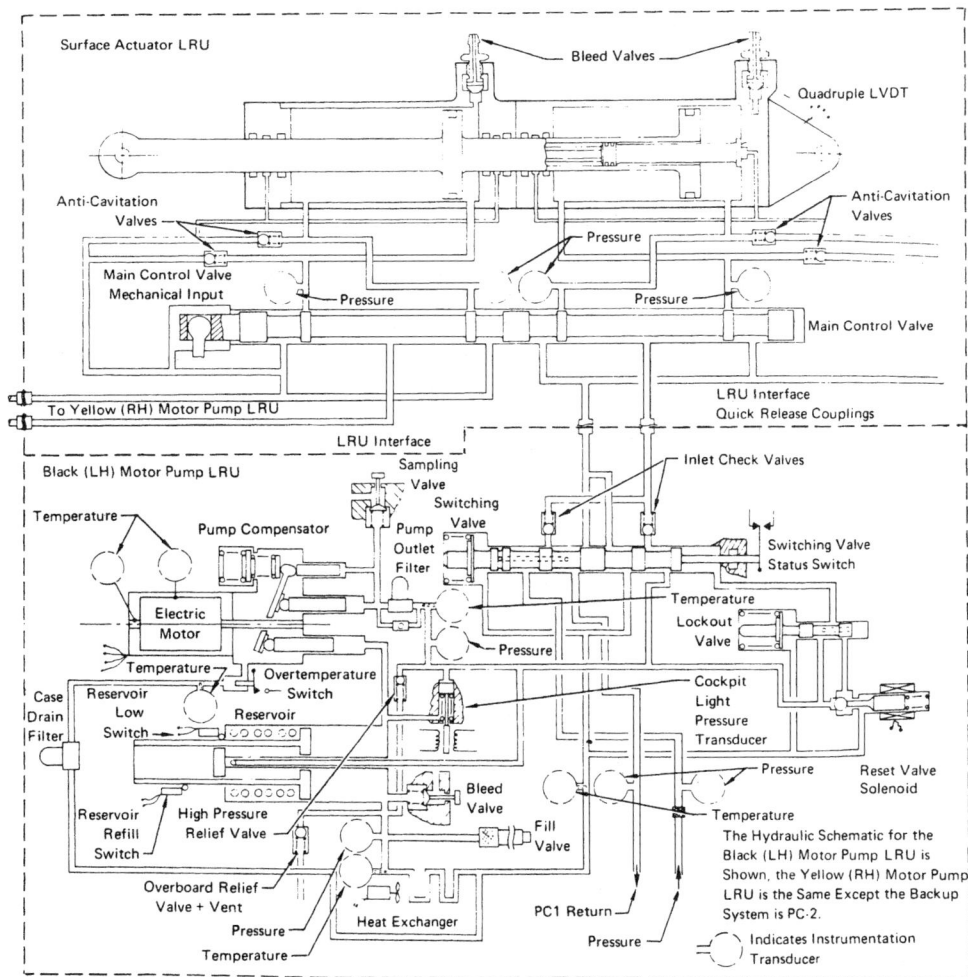

Fig. 55   SSAP hydraulic schematic.

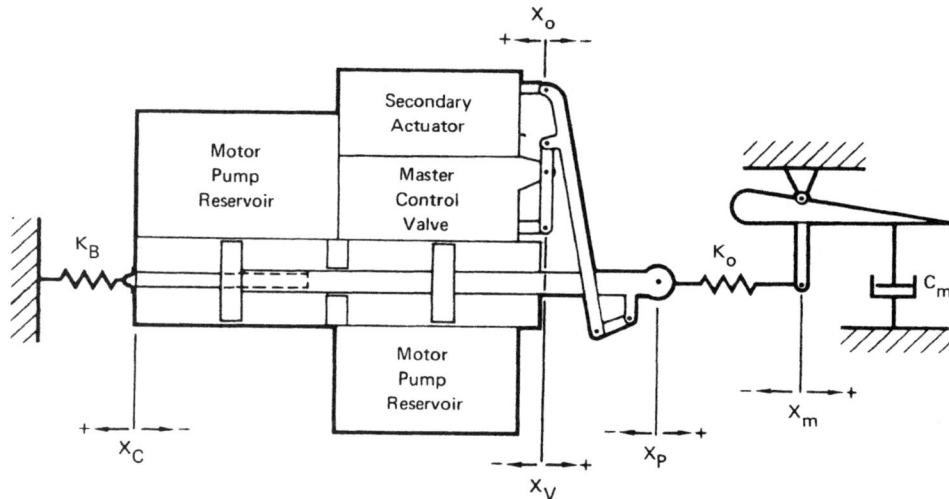

Fig. 56   SSAP schematic.

For the AFFDL contract, LTV developed a duplex design but never tested it beyond the laboratory checkout phase. In regards to parts and components, to achieve a reliability that even approached the standard F-4 installation would require countless hours of testing, not including flight tests, to demonstrate that the new design was flightworthy. Furthermore, this design used an electromechanical secondary actuator that, although analytically proven and laboratory tested, had never been flight qualified. MCAIR's considerable experience during Phase IIA and IIB had been with electrohydraulic secondary actuators. Although the program constraints allowed little time, a compromise design was worked out that added more weight.

As more and more time was consumed and the SSAP's weight increased, it became evident that the lack of sufficient R&D time was going to eliminate any possibility of installing and testing the SSAP. Also, because there had been no previous efforts or similar designs in use, the program faced the problem of redundancy in the flight control actuation area. In many designs the redundancy was put in the outer or aerodynamic loop, making redundant the control surfaces used, each by a single actuator. Again our program constraints prevented any such measures.

Thus, although two SSAP units were fabricated—one intended for laboratory and environmental testing, and the other for installation in the F-4 aircraft for flight tests—neither SSAP was ever installed or flight tested. Phase IIC proved that there were many problems in the area of flight control actuation that needed more R&D work.

Phase IID: Flight Demonstration and Air Force Evaluation

The people responsible for the management of the Survivable Flight Control System (SFCS) program recognized the importance of proving without a doubt that fly-by-wire was a viable option for the design of the next generation of high-performance aircraft. The work under Air Force programs that preceded the SFCS program explored all of the critical issues necessary to make certain the technology was ready. For example, levels of redundancy, probability of catastrophic failure, analog versus digital computer implementation, actuation philosophy, emergency backups, and control law design were among the key technical considerations that had to be addressed. The Air Force Flight Dynamics Laboratory (AFFDL) programs, and especially the in-house fly-by-wire B-47, went a long way toward providing the necessary baseline for the SFCS program. The remaining challenge to the AFFDL people was to close the technological and psychological gap of fly-by-wire. Thus, a major objective of the SFCS program was defined.

The basic Phase II SFCS program was designed to build an experimental package, a high-performance aircraft with an operational-type fly-by-wire system. This meant providing a test bed that could be flight tested by the Air Force people at the controls, who were responsible for setting the requirements and making the decisions on what designs should be used in the next generation of military aircraft. Phases IIA and IIB did just that, yielding a fully production-like operational fly-by-wire system with all the built-in capabilities to test, evaluate, and flight-demonstrate fly-by-wire as a viable, and indeed necessary, design technique for use in flight control systems. Thus, Phase IID was designed to provide for 1) Air Force flight test evaluation, 2) technology transition flights, and 3) demonstration flights. The Phase IID flights were in addition to the contractor development and data flights of Phases IIA and IIB, and they were flown concurrently.

Air Force Flight Test Evaluation

Air Force Phase IID evaluation flight testing was initiated on March 21, 1973 and completed on May 8, 1973. A total of 15 flights were flown jointly by Major R. C. Ettinger of the 4950th Test Wing, ASD, Wright-Patterson AFB, Ohio and Lt. Col. C. W. Powell of the AFFTC, Edwards AFB, California. Testing included evaluations of stability and control, clean and with external tanks, electrical backup control, air-to-air and air-to-ground tracking, gross maneuvering, and precision flying. All flights were flown with the Air Force evaluation pilot in the front seat. Test results are reported in detail in FTC-TR-73-32, "Air Force Evaluation of the Fly-by-Wire Portion of the Survivable Flight Control System Advanced Development Program," August 1973.

A primary objective of this evaluation was to determine the operational suitability of fly-by-wire for inclusion as the primary control system for future aircraft. The Phase IID evaluation reached the following conclusions:

1. Based on this limited evaluation, fly-by-wire was an operationally suitable and ready technology for inclusion in future aircraft.
2. Evaluation of advanced flight control systems, as well as their development and test, should be performed over the full range of the flight envelope, providing that a means of recovery of the aircraft from out-of-control/spin conditions is available.
3. Control stick mechanization using force transducers should be designed to avoid inadvertent torquing inputs, or should be avoided in the basic design in preference to other means such as displacement transducers.
4. A built-in test function proved valuable in diagnosing system status and system failures, and it significantly enhanced safety.
5. The master control and display panel was beneficial to the conduct and safety of the flight test program. Similar, though simplified, displays would be valuable in future operational systems.
6. Flying qualities during take-offs and landings were excellent.

Technology Transition Flights

At the time that the SFCS program was in progress, the Air Force embarked upon what was known as the Light-weight Fighter program. This program was an experiment in prototyping, allowing the contractors wide latitude and flexibility to incorporate and demonstrate advanced technologies. The YF-16 was undertaken by General Dynamics, and the YF-17 by Northrup. Based on the apparent developing success of fly-by-wire for the SFCS program, General Dynamics made the decision to convert the YF-16 mechanical control design to a fly-by-wire system early in the design and fabrication phase. Recognizing that the Air Force Flight Dynamics Laboratory's program was the primary influence for the YF-16 fly-by-wire decision, the SFCS program management very astutely made the decision to maximize the knowledge and experience gained from the SFCS fly-by-wire flight test to be applied to the Air Force Lightweight Fighter program. The result was a block of flights on the SFCS fly-by-wire F-4 to be flown by the pilots of the Lightweight Fighter (LWF) Test Force at Edwards AFB.

Eleven flights were allocated to the LWF Test Force, composed of Lt. Col. Jim Rider, Maj. Walt Hersman, and Maj. M. Clarke. A total of six front-seat and five rear-seat flights were flown with these pilots. Based on this fly-by-wire experience they provided input to the design, development, and test of the YF-16 fly-by-wire flight control system. The pilots' first-hand experience can be credited for the swift success of the YF-16.

Demonstration Flights

Ten demonstration flights were provided to 10 individuals, each of whom was expected to make the appropriate feedback to their organization/unit regarding fly-by-wire. Most of the demo pilots had never flown an F-4, and a number of them were not fighter pilots. In most instances, their adaptation to some of the unique features of the SFCS fly-by-wire system such as the side-stick controller was much quicker than that of experienced F-4 pilots. A partial explanation for this was that they did not have to go through the unlearning process of how to fly a conventional mechanically controlled F-4 before concentrating on flying the SFCS fly-by-wire F-4.

The pilots flying the demo flights were:

> Maj. Gen. V. Turner, Chief of Staff, Air Force Systems Command
> Capt. L. A. Walker, United States Marine Corp.
> G. Krier, Test Pilot, National Aeronautics and Space Administration
> Lt. Col. Babin, B-1 SPO, Aeronautical Systems Division
> Maj. Johns, 4950th Test Wing, Aeronautical Systems Division
> Maj. R. Barlow, AF Test Pilots School, Edwards AFB
> Maj. H. Dibble, Tactical Air Command, TAWC, Eglin AFB
> Maj. I. Payne, 4950th Test Wing
> Maj. D. Cairns, 4950th Test Wing
> Maj. R. Nadig, Safety Office, Aeronautical Systems Division

The following were some representative comments from the pilots' postflight briefings:

> "Even though I'm not an F-4 pilot, the aircraft was easy to fly."

> "The aircraft was very well damped."

> "Formation flying was very stable and *g* tracking extremely easy."

> "BIT was very impressive."

> "Transonic deceleration was a definite improvement over the basic F-4."

> "Handling qualities are very good."

> "Lateral axis very sensitive."

> "After three or four flights, a pilot can become very proficient using the side-stick controller."

> "I thought the flight was tremendous."

"Damping in all three axes was dead-beat."

"If the roll axis were optimized to the state of the pitch axis, it would be a great flight control system."

"Air-to-air tracking was very good, and better than the simulator."

"The thing that impressed me the most is that you don't think anything about the system being so unique. You just start flying the aircraft and do what you want to without worrying about it."

Phase IIE: Precision Aircraft Control Technology and Reliability/Maintainability of SFCS Fly-by-Wire

Perhaps one of the first real attempts to exploit and prove the beneficial application of fly-by-wire was the Precision Aircraft Control Technology (PACT) program. The Air Force Flight Dynamics Laboratory SFCS program people very creatively molded a program to extend a McDonnell Aircraft Co. IRAD[5] program to flight test a canard configured F-4 to exploit the real benefits of active control. The aircraft chosen for this experiment was the SFCS fly-by-wire equipped F-4 because of its proven reliability and performance. At the same time, this program would provide additional reliability and maintainability data based on actual operational conditions, thus closing the technological and psychological gap once and for all.

Thus, Phase IIE was born. The primary objective of the effort was to demonstrate the use of active control technology as an integral design parameter in evolving aircraft configurations. This was successfully done. Most recent examples where this paid off are the unique configurations of the AFTI F-16 and the forward swept wing X-29 in the R&D community and the F-117 and B-2 in the operational world. It can be stated unequivocally that the success of these examples can be credited to the SFCS and PACT work and to all of the Air Force flight control research and development tracing back to the mid-1950s.

The PACT program used the SFCS F-4 with a modified control system, and added horizontal canards and the capability to manage internal fuel to enable flight with negative static longitudinal stability margins up to $-7.5\%c$. The canards were designed with differential deflection capability for side-force generation, although this feature was not demonstrated in the PACT program. They were operated symmetrically for PACT and used a dual-channel electronic system with two hydraulic systems to drive the surface actuators. The canard location and size were based on various designs tested in the wind tunnel.

A total of 34 flights were flown. On completion of Phase IID, the SFCS F-4 was put in storage on June 3, 1973. When it was reactivated some six months later for the PACT program, the SFCS fly-by-wire system had passed its initial built-in tests with flying colors. The data flights were flown primarily by McDonnell Aircraft Co. pilots. However, in keeping with the philosophy of emphasis on technology transition and application and closing the fly-by-wire technological and psychological gap, the flights incorporated demonstrations to the following pilots:

Rear Admiral F.H. Foxgrover, HQ Naval Air Systems Command
Capt. L. A. Walker, USMC, Naval Air Test Center
Maj. R. C. Ettinger, Lightweight Fighter Test Force
Col. B. D. Ward, Commander, Air Force Flight Dynamics Laboratory

---

5. In-house Research and Development.

Phase IIE flights successfully explored the benefits of fly-by-wire to active control technology, multimode controls for mission segment optimization, blending of surface controls and relaxed static stability to achieve improved performance, and CCV (control configured vehicles) as a key design tool for future aircraft. Above all, this phase demonstrated unequivocally that fly-by-wire can be implemented to provide a totally safe and reliable flight control system with a high degree of confidence, and with the benefit of significantly fewer maintenance man-hours per flight than with conventional mechanical flight control systems.

Chapter 5

Technology Transition and Application

Fly-by-wire is a flight control system wherein vehicle control input is transmitted completely by electrical means. As discussed earlier, there were any number of control systems that, in one form or another, transmitted some control signals via electrical wires. Autopilots and missile controls of the 1950s are probably the most notable and well-understood examples. However, given their short intended operational life and the fact that human life did not depend on their reliability and integrity, they did not really qualify as the kind of control required for a manned aircraft application. Since the early 1950s, flight control engineers of Wright-Patterson Air Force Base's Flight Dynamics Laboratory, and its preceding organizations, envisioned fly-by-wire as a major enabling technology vital to the development of future Air Force airplanes. In addition to design simplification, fly-by-wire would pave the way for precision control of the total aircraft flight path and for high levels of system integration.

The Air Force Flight Dynamics Laboratory pioneered fly-by-wire by a series of R&D exploratory programs that ultimately provided the technology to design the systems we have in use today. In the 1960s a number of studies were undertaken, including the B-47 in-house program, to establish the technology for system mechanization by designing, fabricating, testing, and flight-testing flightworthy hardware. In addition, control law developments for precision maneuver and flight path response control were being validated through flight tests on the USAF's F-102 test aircraft and the F-4 TWeaD airplane. These key forerunners, which included some industry mechanization, paved the way in aviation history for fly-by-wire manned aircraft.

But these efforts were not enough. The skeptics prevailed. They believed it impossible to build a safe and reliable flight control system solely dependent on electrical command signaling. The U.S. government agencies responsible for acquiring new airplanes would not think of committing to a new major weapon system acquisition that relied on such high-risk technological innovations. Thus, the major airframe contractors would not propose such advanced technology in their proposals to the Air Force for fear that they would be ruled out of the competition. Test pilots and operational command pilots could not imagine an airplane without mechanical connections between their control stick and the airplane's control surfaces. The main dilemma and challenge then was the need to close the technological and psychological gap of fly-by-wire.

On December 16–17, 1968, the Air Force Flight Dynamics Laboratory (AFFDL) held a fly-by-wire conference at Wright-Patterson AFB in Ohio. The purpose of this conference was to bring the industry and government agencies together to share in their views about fly-by-wire, its potential, actual experience, and what they felt needed to be done to close the psychological and technological gap. The results of the AFFDL fly-by-wire programs performed by Sperry Phoenix and Douglas Aircraft Long Beach were presented by their respective project engineers. The AFFDL reported on exceptional results of the B-47 in-house fly-by-wire for advanced fighters (e.g., the yet-to-come F-15), large strategic bombers such as the AMSA (later to become the B-1), and helicopters. Philosophy of redundancy, digital computation, handling qualities criteria, and pilot acceptance were topics of round table discussion. Individual conclusions indicated that, although fly-by-wire systems offer many advantages over their mechanical counterparts, government-sponsored flight tests would be required before the confidence level in such a system would be high enough to permit its general acceptance.

Concurrently with promoting R&D for technology, the late 1960s saw the AFFDL focus on the immediate need to support the Air Force effort in Vietnam. Studies were conducted both in-house and by contract using real experience and data from Southeast Asia. It was determined that aircraft flight control system combat survivability could be enhanced significantly by incorporating redundancy both in signal transmission and actuation. Fly-by-wire was made to order. Fly-by-wire had found the technological pull it needed to mature its development for future implementation under the 680J Survivable Flight Control System Program. This program culminated in the world's first fully operational fly-by-wire system to be test flown successfully again and again on a high-performance manned aircraft, the AFFDL's F-4 S/N 62-12200.

The success of the 680J fly-by-wire development program in the early 1970s, and its associated evaluation and demonstration flights, provided General Dynamics (GD) with a viable and necessary flight control technology for the YF-16 being developed for the Air Force's Lightweight Fighter program. Harry Hillaker of GD (father of the F-16) was quoted as saying that "if it were not for the Air Force Flight Dynamics Laboratory's pioneering fly-by-wire work, there would be no F-16 as we know it today."

The YF-16 was the single-engine entry in the lightweight fighter prototype project. Fly-by-wire provided the GD design with improvements in handling qualities, precision, and responsive control for target tracking, increased reliability, less maintenance, and an aerodynamically unstable airplane that could fly circles around any of its heretofore competitors. The YF-16 won the down selection largely due to the benefits gained from fly-by-wire, and it became the world's first fleet aircraft to be procured that implemented this radical new technology called fly-by-wire.

The YF-16 competitor was the YF-17 provided by Northrup. Upon selection of the YF-16 by the Air Force, Northrop teamed up with McDonnell Aircraft to redesign the YF-17 to incorporate a digital fly-by-wire system. The McDonnell engineers who did this work were the same ones who had earlier done the 680J fly-by-wire program for the AFFDL. The fly-by-wire system implemented was essentially a repeat of the 680J system, except that the Navy required a mechanical backup system be provided. Resulting from all this was the F/A-18 procured by the Navy and Marine Corps.

As the military aerospace industry began adopting fly-by-wire to reap its manufacturing, performance, and marketing benefits, AFFDL continued to experiment using the YF-16 to demonstrate fly-by-wire applications for maneuver enhancement, direct lift and side-force control, and variable relaxed static stability. This program was known as the Control Configured Vehicle (CCV) program. AFFDL also developed digital fly-by-wire technology and demonstrated it on its A-7 Digitac flight test vehicle. Digital computation capability was exploited to improve fault tolerance and redundancy management. Beneficiaries of the CCV and Digitac programs were the X-29 Forward Swept Wing experimental airplane and the Advanced Fighter Technology Integration (AFTI) program.

Through the use of digital fly-by-wire, AFFDL proved that unorthodox, statically unstable designs like the X-29 could be safely flown. The AFTI program was an extensive R&D effort to address integration of the fly-by-wire control system with other essential subsystems, such as avionics and stores management to improve the capability of the weapon platform. By the mid 1980s, fly-by-wire had become the industry standard for flight control systems and was being implemented on virtually all new Air Force aircraft acquisitions.

The technical knowledge gained from the U.S. Air Force's fly-by-wire work has led to unique airframe designs to meet specific mission and force structure requirements. Two good examples are the F-117 and the B-2 stealth aircraft. Fly-by-wire provides the airframe designer with the capability to develop low radar signature aircraft without major impacts to flight control system complexity and flight safety. Both aircraft are statically unstable, and now fly-by-wire makes it possible to design such configurations. The B-2 and the F-117 fly-by-wire systems are integrated with weapon delivery to provide for precision launch/drop capability, and with the avionics systems to take advantage of digital mapping, terrain following, and mission planning. Other systems using fly-by-wire for its obvious advantages are the C-17, the advanced aerial refueling boom for the KC-10, the F-15E, the F-22, and the Joint Strike Fighter.

Success of fly-by-wire for the U.S. Air Force has also led to foreign applications such as the Saab JAS-39 and the India Light Combat Aircraft. Additionally, in the 1980s, Airbus Industries introduced fly-by-wire aircraft into the commercial sector as a means to differentiate its product in both performance and cost. Airbus realized the same benefits as had the U.S. Air Force in adapting fly-by-wire, and it reduced its customers' overall cost of aircraft ownership by offering them the A-320 Airbus, which is more reliable, easier to maintain, and safer to fly. In the domestic aircraft market, Boeing also identified and concurred with fly-by-wire benefits and incorporated the technology in its Boeing 777 aircraft.

Closing Thoughts

The most important aspect of fly-by-wire development and implementation is the role the U.S. Air Force played in technology innovation. The research and development information resulting from the work conducted by AFFDL and its predecessor organizations through in-house and contract efforts was given the widest distribution appropriate. No one contractor walked away from AFFDL projects with a competitive technology edge over the others. For example, McDonnell Aircraft was the AFFDL contractor that developed and tested fly-by-wire for the USAF under the 680J program. But the first airframer to incorporate the technology was General Dynamics for the F-16. Another example is the X-29 program in which Grumman Aerospace used digital fly-by-wire for precise and safe control of the uniquely unorthodox airframe. Northrop gained its experience with fly-by-wire through the F/A-18 and the B-2. Lockheed Corporation implemented fly-by-wire in its Skunk Works project F-117. Douglas Aircraft implemented fly-by-wire control for its aerial refueling boom. AFFDL assured technology availability to any system developer who chose to benefit from it.

By the 1980s, AFFDL flight control engineers were world renowned for their pioneering work in fly-by-wire. They spread the gospel through symposia, seminars, information releases, and special cooperative ventures. The net result has been to open up new fly-by-wire markets for U.S. firms that specialize in control science technology. Indeed, AFFDL is credited for having closed the psychological and technological gap of fly-by-wire for the U.S. Air Force. In military terms, our Air Force was provided with superior aircraft for combat in the event of hostile action by any adversary.

References

1. AFFDL. *Proceedings of the Fly-by-Wire Flight Control System Conference.* December 1968. AFFDL-TR-69-58.

2. Amies, G., C. Clark, C. Jones, and M. Smyth. *Survivable Flight Control System Studies, Analyses, Approach.* May 1971. AFFDL-TR-71-20 Supplement No. 3.

3. Bazill, D., and G. Jenney. *Research on Flight Control Systems, Vol. III: Fly-by-Wire Techniques.* October 1970. AFFDL-TR-69-119 Vol. III.

4. Emfinger, J. E. *A Prototype Fly-by-Wire Flight Control System.* August 1969. AFFDL-TR-69-9.

5. Hooker, D., R. Kisslinger, G. Smith, and M. Smyth. *Survivable Flight Control System, Interim Report No. 1.* May 1971. AFFDL-TR-71-20.

6. Hooker, D., R. Kisslinger, G. Smith, and M. Smyth. *Survivable Flight Control System Studies, Analyses, Approach.* May 1971. AFFDL-TR-71-20 Supplement No. 1.

7. Hooker, D., R. Kisslinger, G. Smith, and M. Smyth. *Survivable Flight Control System, Interim Report No. 1. Studies, Analyses and Approach.* May 1971. AFFDL-TR-71-20 Supplement No. 2.

8. Hooker, D., R. Kisslinger, G. Smith, and M. Smyth. *Survivable Flight Control System Final Report.* December 1973. AFFDL-TR-73-105.

9. Jenney, Gavin D. *JB-47-E Fly-by-Wire Flight Test Program Phase 1.* September 1969. AFFDL-TR-69-40.

10. Jenney, Gavin D. *Research on Flight Control Systems Fly-by-Wire B-47 Phases II & III.* August 1970. AFFDL-TR-69-119.

11. Koch, W. G. *Research and Development of an Integrated Servo Actuator Package for Fighter Aircraft.* November 1969. AFFDL-TR-69-109.

12. Larson, H., and C. Zimmer. *Military Transport (C-141) Fly-by-Wire Program.* April 1974. AFFDL-TR-74-52.

13. Miller, F., and J. Emfinger. *Fly-by-Wire Techniques.* July 1967. AFFDL-TR-67-53.

14. Osder, Stephen, and David LeFebre. *Modification of Prototype Fly-by-Wire System to Investigate Fiber-Optic Multiplexed Signal Transmission Techniques.* March 1974. AFFDL-TR-74-10.

15. Powell, C., Lt. Col. *Air Force Evaluation of the Fly-By-Wire Portion of the Survivable Flight Control System Advanced Development Program.* August 1973. FTC-TR-73-32.

16. Sethre, V., R. Hupp, and G. Rayburn. *Design and Evaluation of a Single Axis Redundant Fly-By-Wire System.* December 1968. AFFDL-TR-68-81.

17. *Sperry A-12 Gyropilot.* Publication No. 15-130. Sperry Gyroscope Co., Inc. Great Neck, NY: February 1946.

Appendix A

Survivable Controls Gain Emphasis

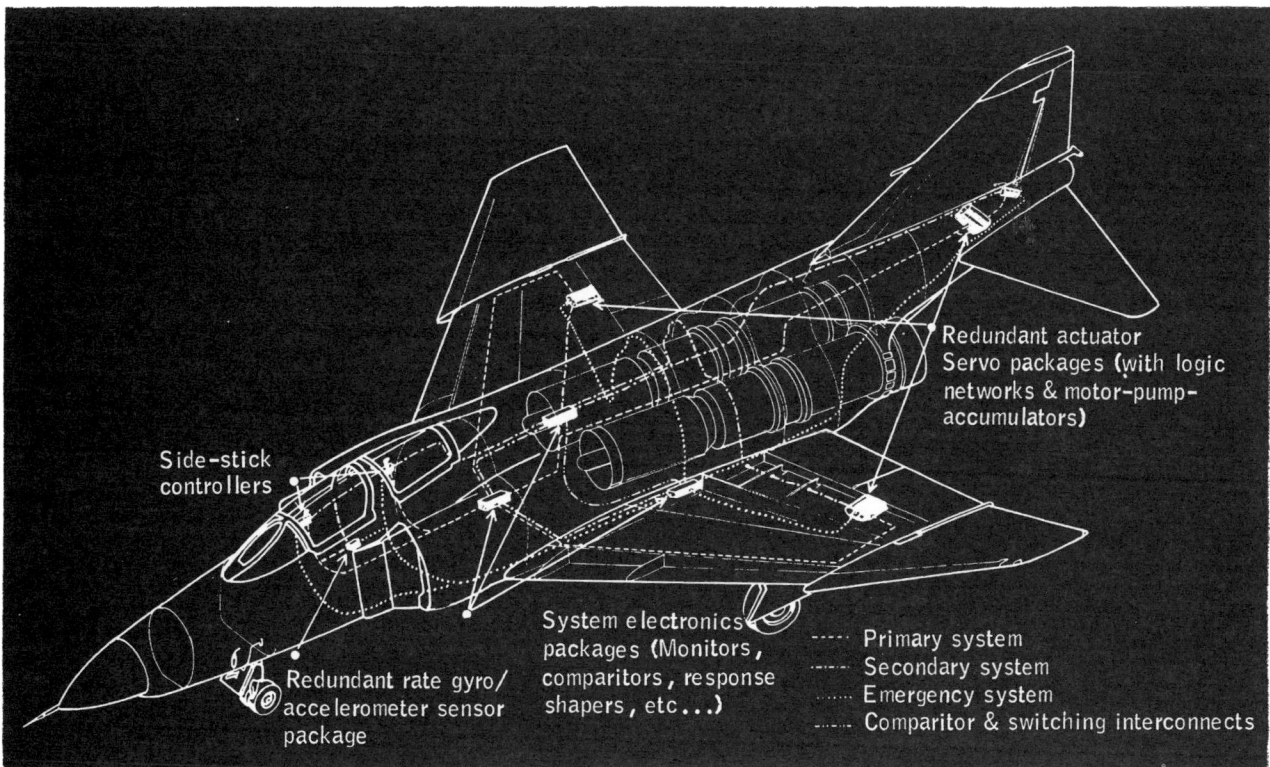

Survivable flight control system layout is shown as it might be installed in a McDonnell Douglas F-4. System is designed to with-stand two similar failures and still leave pilot with sufficient control to return to his base and land the aircraft.

Avionics

Survivable Controls Gain Emphasis

USAF plans to eliminate long hydraulic lines, mechanical links by use of redundant dispersed wiring, self-contained actuators

By Michael L. Yaffee

Wright-Patterson AFB, Ohio—USAF Flight Dynamics Laboratory is developing its survivable flight control system (SFCS) effort at a pace that could provide at least a limited payoff in survivability and improved operating performance for early models of aircraft now on the drawing boards, such as the F-15 (AW&ST Jan. 19, p. 22).

This system, which combines fly-by-wire aircraft control with integrated hydraulic servo-actuator packages, is based on the idea that dispersed redundant control elements will reduce vulnerability to enemy firepower, improve control perform-ance and increase stability of the aircraft as a weapons delivery platform.

Elements of such a system are already in operation in other countries on both commercial and military aircraft. The Flight Dynamics Laboratory and Mc-Donnell Douglas Aircraft Co., which has received a 44-month $16.5-million USAF contract for this effort, are trying to develop a new and more extensive survivable system for U.S. aircraft—both military and commercial.

A McDonnell Douglas F-4 will be used as a flight test vehicle, at least early in the program (AW&ST Aug. 4, 1968, p. 113).

Conceivably, an early, relatively simple form of this system could be available for new or existing F-4s. But the Air Force, which funded some of the early actuator work out of its F-15 system program office, considers this unlikely now. More probable recipients of the benefits of survivable flight con-trol systems will be new aircraft such as the F-14, F-15 and B-1.

Purpose of the fly-by-wire program is to eliminate the long hydraulic lines and mechanical linkages that run through the aircraft to connect the pilot's control stick to the actuators on aircraft control surfaces. These would be replaced with redundant, dispersed wires that will carry signals from the pilot to self-con-tained actuator packages.

The idea is that the new survivable flight control system—which would actually be a combination of fly-by-wire and power-by-wire (the self-contained actuator packages) systems—would be significantly less vulnerable to enemy fire. There would be side benefits such as greater weapons delivery platform stability, but greatly enhanced survi-vability is the prime current objective of the Air Force program.

Under its contract with Flight Dy-namics Laboratory, McDonnell Doug-las will install and fly a self-contained or integrated actuator package (IAP) in an F-4 in March or April. This IAP, which was designed and fabricated by LTV Electrosystems under contract from the laboratory, will be used to control the F-4 stabilator and, thereby, the motion of the aircraft about the longitudinal or pitch axis.

The LTV package is a simplex unit, somewhat like the standard F-4 stabila-tor actuator it will replace. It is designed to operate with the aircraft's present hydraulic systems. It differs primarily in that it also has a self-contained hydraulic system—fluid, reservoir, electric motor and pump—that will enable the pilot to operate the stabilator for as long as 2 hr. if the main hydraulic system is shot away or fails for some other reason. In an emergency, the simplex unit would cut in automatically. For test purposes, the pilot will be able to switch the unit into its automatic operational mode.

Once activated, the simplex emergency power system can move from dead stop to full power in less than ½ sec. In addition to providing sufficient control capability for ordinary maneuvering of the aircraft in flight, according to Flight Dynamics Laboratory's Vernon R. Schmitt and Capt. Robert C. Lorenzetti, the simplex package will have sufficient flow and pressure to enable the pilot to land the aircraft.

Power inputs to the simplex package are electrical when the unit is operating independently of the aircraft's hydraulic system. This is the basis for the term power-by-wire system which is being applied to these integrated actuator package concepts.

On completion of the flight test program, the LTV simplex package will be replaced with the regular F-4 stabilator actuator. This will end Phase 1 of the laboratory's survivable flight control system program.

Next, in Phase 2A, McDonnell Douglas will install and test fly a quadruple redundant fly-by-wire system in the F-4. The company issued proposal requests in December to outside contractors who would actually build the system.

The system will be a closed-loop electro-hydraulic feedback control system operated by a small limited-motion sidestick built into a pilot armrest. Instead of an elaborate feel system used with conventional mechanical linkages and cables, the fly-by-wire control system will use linear springs and will control aircraft motion rather than control surface position.

The Phase 2A system in the F-4 will be what is termed a pseudo fly-by-wire system since the aircraft will retain its regular, mechanical control system. Although disengaged during flight tests of the fly-by-wire system, the mechanical system will be available for emergency use. Flight testing is scheduled to begin in September, 1971.

In Phase 2B of the program, beginning in January, 1972, the mechanical control system will be removed from the F-4. The aircraft will be flown on a true fly-by-wire electrical control system. There will be no continuous mechanical linkage between the pilot's control stick and the control surface actuators. The objective here is to demonstrate the confidence of Flight Dynamics Laboratory and McDonnell Douglas engineers in a pure fly-by-wire control system.

The quadruple redundant fly-by-wire

system with its four signal paths, four analog computers, four sets of motion sensors and four transducers was chosen to provide the desired level of reliability.

A quadruple electrical system, Lorenzetti said, can withstand two similar failures—two wires or two transducers, for example—and still operate normally.

The four channel, electrical, fly-by-wire system will provide reliability at least as high as that offered by the present mechanical control system in the F-4. Also, although it would be possible to go to a quintuple redundant or even higher, the quadruple system provides the best payoff, according to Air Force and McDonnell Douglas calculations. In the event of a third similar failure in the quadruple redundant fly-by-wire system, Lorenzetti said, the system will go to a neutral or preselected trim position.

In the final part of the program, Phase 2C, the Air Force Flight Dynamics Laboratory and McDonnell Douglas will combine the fly-by-wire system with a power-by-wire system to get a survivable flight control system that will be flown in the latter half of 1972. Originally, plans called for use of integrated actua-

tor packages (power-by-wire) in the longitudinal and lateral axes of the aircraft. For reasons of economy, present plans now limit installation of the actuator packages in the longitudinal axis. Fly-by-wire control will be used in all three axes.

The fly-by-wire system used in Phase 2C will be the same one used in Phases 2A and 2B. About five companies—not necessarily the same five—are expected to bid for both the fly-by-wire and the power-by-wire contracts. The integrated actuator package that will be used in Phase 2C is expected to be similar to the duplex integrated actuator packages developed by LTV and General Electric under contracts from the Flight Dynamics Laboratory. There is also a possibility that the actuator might be triplex unit.

The duplex integrated actuator packages built by LTV and GE are designed to operate as completely self-contained actuators in conjunction with fly-by-wire electrical control systems. These two units, however, are non-flight-rated, prototype hardware.

The original program called for LTV to build a non-flight rated simplex and

Simplex integrated actuator package (above) and duplex integrated actuator (below) were built by LTV Electrosystems under a USAF Flight Dynamics Laboratory contract. The simplex unit will be flight tested on a McDonnell Douglas F-4 stabilator in March or April. General Electric has also built simplex and duplex integrated actuator packages, but both companies' duplex units are non-flight-rated hardware. The laboratory plans flight tests of a completely self-contained duplex or triplex actuator for a fly-by-wire electrical control system in 1972.

a duplex package and to provide the Air Force with a design—but no hardware—for a triplex actuator package. About three and a half months into the design effort, USAF decided it would fly a simplex integrated actuator package.

In order to have a backup for the new flight program, the laboratory issued the same work statement to a second contractor, General Electric's Aircraft Equipment Div. Like LTV, GE built a flight-rated simplex package and a non-flight-rated duplex unit sized for application to an F-4 stabilator. GE also developed a design for a triplex actuator. To save money, the laboratory deleted the triplex requirement from LTV's work statement. GE started later than LTV and stayed with the cast aluminum body like the one used in the regular F-4 stabilator actuator, while LTV built a stainless steel actuator body.

The laboratory selected the LTV simplex package for flight test on the basis of its lighter weight and superior resistance to crack propagation. Sperry Rand's Vickers Div. built the pumps for the LTV and GE simplex units and the LTV duplex unit. A GE pump is used in the GE duplex package.

Flight Dynamics Laboratory and McDonnell Douglas will decide soon between a duplex unit and a triplex unit for Phase 2C of the program. As opposed to the simplex actuator, which has a limited capability for self-sustaining operation, the duplex and triplex units are designed to operate full-time completely independently of central hydraulic systems with their exposed plumbing, Schmitt said. The duplex unit has two independent motor-pump-reservoir units that power a dual-tandem or four barrel ram. Either half of the duplex unit can provide full-rated, hinge moment and surface rate. The triplex design calls for three independent motor-pump-reservoir units.

Development and use of a closed-loop electrical control system in future aircraft could be justified on other grounds than enhanced survivability, according to Lorenzetti and Schmitt. A significant improvement in overall aircraft performance is an inherent benefit in closed-loop control systems such as fly-by-wire, which can function both as a stability augmentation system and a control augmentation system. Specific benefits they foresee include:

■ Greater versatility than conventional hydro-mechanical control systems. A survivable flight control system or just a straight fly-by-wire system, they said, would permit a pilot to operate his ailerons differentially to get roll control, as he now does with mechanical linkages. In addition, by changing the signs of his electrical inputs or voltages, he could operate his ailerons together to get pitch control. This increased versatility also leads to increased flexibility which, in turn, further enhances survivability. If the stabilator became inoperative, the pilot could obtain limited pitch control with ailerons.

■ Improved weapons delivery would reduce the number of passes over a target and, hence, exposure to enemy fire. In a related Air Force program, tactical weapon delivery (TWeaD), the Air Force learned that providing a pilot with a precision vernier control capability that enables him to make very small precise changes in his flight path makes it possible for him to line up on his targets faster and more accurately. With a fly-by-wire control system already in the aircraft, Lorenzetti and Schmitt said, it becomes possible to add systems such as vernier control at minimum or negligible costs.

■ Tailoring airplane control to mission phases. In current aircraft, Lorenzetti said, the designer and pilot generally have to accept performance compromises. Optimum landing performance, for example, may be sacrificed for satisfactory cruise handling and both of these may be compromised for suboptimum but adequate handling during weapons delivery. With a closed loop electrical control system, these compromises can be eliminated or significantly reduced, according to Lorenzetti and Schmitt.

■ Weight reduction. Replacing hydraulic lines, steel rods and linkages, pulleys and cables with electrical wires and self-contained integrated actuator packages should lead to lower weights.

Availability of a proved fly-by-wire system, Lorenzetti said, will make possible even larger weight reductions through a basic change in aircraft design philosophy. Along this line, the laboratory is proposing an advanced development program on a "controls configured vehicle." This program

would be concerned with designing a new aircraft around a fly-by-wire system with the goal of saving weight. Today's aircraft that have stability augmentation systems to enhance their inherent configuration stability still require enough basic stability to enable the pilot to fly the aircraft if the stability augmentation system becomes inoperative.

This stability margin is generally bought by the pound in the form of sheet metal needed to build big vertical stabilizers and the like. A fly-by-wire system can perform the functions of a stability augmentation system with sufficient additional reliability to eliminate the need for larger stabilizer surfaces, according to Lorenzetti and Schmitt.

■ Improved ride, increased fatigue life, reduced structural weight. All these can be achieved through addition of load alleviation and mode stabilization (LAMS) inputs to the basic fly-by-wire system.

■ Reduced pilot workload. In an electrical control system with a side stick working against adjustable feel springs, the pilot can keep his arm on an armrest and control an airplane with comparatively slight finger or hand motions regardless of turbulence or high gravity maneuvering forces. With a conventional, center control stick, mechanical linkages and long heavy control cables, the pilot has to work much harder in a high force field just trying to hold the control stick in a neutral position.

The benefits offered by systems such as TWeaD, LAMS, stability augmentation system, control augmentation system, direct lift control, optimal automatic terrain following and even the automatic pilot system can be obtained more simply and readily by building upon the basic fly-by-wire part of a survivable flight control system, Lorenzetti said.

In fact, he added, load alleviation, gust alleviation and complexly coupled control systems such as those used or planned for VTOL aircraft and manned lifting, re-entry vehicles rely on motion feedbacks and thus require fly-by-wire to realize their full potential, without the compromises associated with hydro-mechanical systems.

Appendix B

F-4 Fly-by-Wire . . .
Research Platform for the Future

product support digest

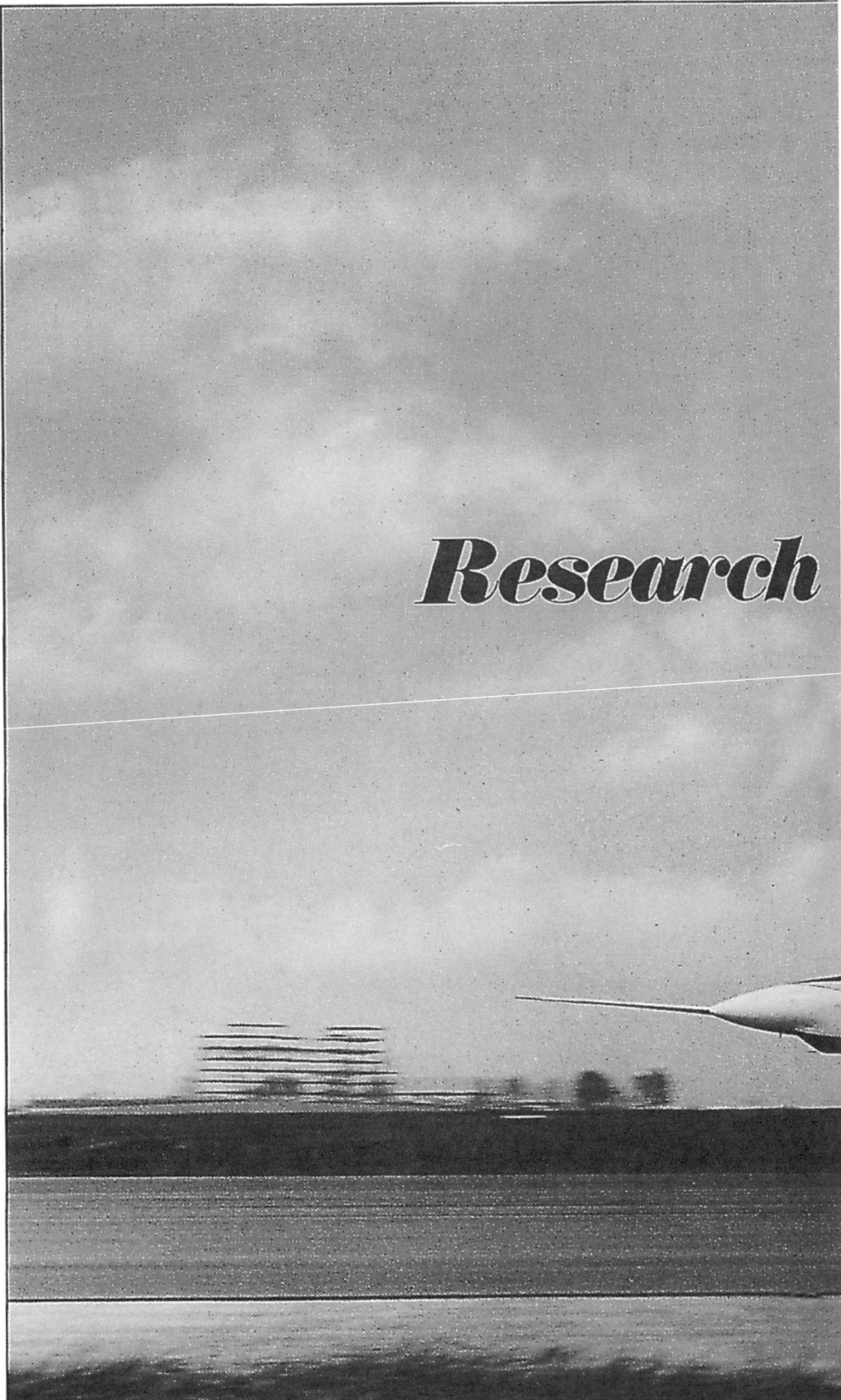

product support *digest*

A System for Tomorrow... Today!

JOHN F. SUTHERLAND / Vice President, Product Support; JOHN J. PETERSEN / Director, Product Service; VERNON E. TEIG / Director, Support Operations.

DIGEST STAFF - EDITORIAL: Editor/Nade Peters; Associate Editor/John Waidmann; Staff Editor/Bruce Mitchell; Editorial Assistant/Cathy Boevingloh. TECHNICAL ADVICE: Product Service Specialist Group. ART & PRODUCTION: Graphic Arts.

VOLUME 19 3RD QUARTER 1972

Research

FLY-BY-WIRE

Reprinted with permission from *Product Support Digest.*

F-4
Fly-by-Wire...
Platform for the Future

On April 29th of this year, a sleek blue and white Phantom aircraft
took off from St. Louis International Airport to become the first U.S. high-performance
jet fighter to maneuver with a computer-controlled "Fly-by-Wire" system.

The Fly-by-Wire test flights - now transferred to Edwards Air Force Base in California - are part of a SURVIVABLE FLIGHT CONTROL SYSTEM research and development program directed by the U.S. Air Force Flight Dynamics Laboratory at Wright-Patterson Air Force Base, Ohio. USAF Program Manager is James W. Morris. McDonnell Engineering Manager is David S. Hooker. Pilot of the test aircraft is Charles P. "Pete" Garrison, McDonnell Project Experimental Test Pilot.

Although the primary purpose of this R&D program is to develop systems which will help the aircrew and aircraft survive in combat, better handling and more accurate weapons delivery are also goals. And it also promises such future advantages as lightweight, compact, highly redundant, more reliable electronic flight control systems. These, in turn, can lead to radical changes in airframe design - changes such as smaller control and stabilizing surfaces, lighter-weight wings, and less structural weight.

The key to the future of Fly-by-Wire lies in answering the basic question of whether AVIONICS control systems can be made as reliable as MECHANICAL control systems. This Special DIGEST section - featuring the viewpoints of the pilot, the engineer, and the maintenance man - describes how McDonnell is helping the U.S. Air Force answer that question! Even though you pilots will not be climbing into the cockpit of an FBW Phantom tomorrow, or you ground technicians facing the immediate prospect of FBW system maintenance, we think you'll all be mighty interested in this account of what's around the next corner of fighter aircraft development.

Why Fly-by-Wire ?

By PETE GARRISON / *Project Experimental Test Pilot*

Once again, I'm caught trying to meet an editorial deadline on a subject which requires much time and thought to properly get on paper. Fortunately, the nuts and bolts of this F-4 Survivable Flight Control System are nicely covered in companion articles in this issue. Thanks to the efforts of McDonnell's Project Engineering and Maintenance Engineering people, I have some license to philosophize a bit about the concept of Fly-by-Wire from a pilot's point of view. A line-up of potential controversial areas would include some whys, whats, and whens.

THE WHY

My personal observations about the subject of Fly-by-Wire have ranged from a beginning of "Who needs it?" to a present "I think it's the way to go!" First of all, we have come to what I see as the crossroads in the fighter business. If we are to continue the manned fighter, we need a significant increase in performance. Even with present structural technology, we are approaching a point of diminishing returns in

trying to give the fighter pilot more favorable power loading and wing loading. These parameters are two of the most important, if not the most important, in the fighter-versus-fighter arena.

Specifically, we need a major breakthrough in the reduction of structural weight in fighter aircraft. One method of doing this is reduction of the vehicle stability margin - i.e., to fly the aircraft with the center of gravity at or aft of the neutral stability point (see diagram). This reduces the structural strength required of the aft fuselage and also reduces the size of the control surface needed to do the job.

The same general philosophy can be applied to the directional axis. What evolves is a control-configured vehicle versus a stability-configured vehicle . The obvious question then becomes, "If neutral or unstable aircraft are so great, why don't we build them that way?" Simple: The human animal's beautifully designed analog computer can't generally cope with sensitive, high-frequency, unstable motion re-

sponse. Remember that last PIO? Therefore, a method of controlled motion response is required.

However, I suspect that every pilot who has strapped on a high-performance fighter has an instinctive distrust of electronic gadgetry. This defensive attitude usually stems from one or more unpleasant personal encounters with various autopilots and/or stability augmentation systems which, without warning, went unstable or hardover; that is, "the damn thing tried to bite!" Since many of those early electronic systems were designed for pilot relief or just a bit of flying quality fine tuning, the stalwart hero of "fist over gadgets" simply turned it off and went his merry way, firmly convinced that he was still master of the valley. Obviously, reliability must be of prime concern.

With the age of high-speed airborne computers and associated electronic technology, we have within our grasp potential answers to the most important questions of reliability and survivability and controlled motion response.

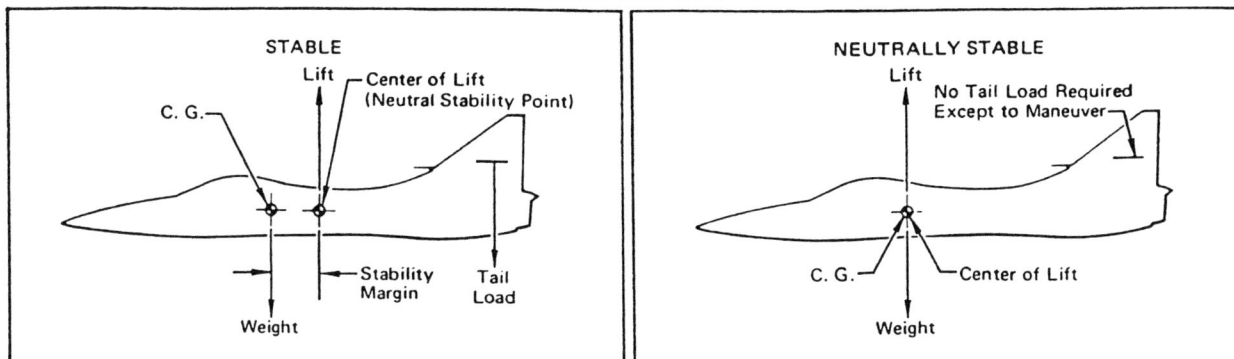

VEHICLE STABILITY

THE WHAT

The concept of a Survivable Flight Control System (SFCS) has long been the aim of the aerospace industry and the U.S. Air Force. The awarding of a contract by the Air Force Flight Dynamics Laboratory to McDonnell Douglas Corporation to design and build such a system in an F-4 aircraft is a concentrated effort to reduce concepts to flying hardware. Since the SFCS is designed to sustain a high degree of damage and still continue to function, it is by design a highly reliable redundant system. In addition, since it is a motion command system with feedback response, the potential exists for control of either stable or unstable aerodynamic vehicles. As you can see, I have very cleverly led you to conclude that a Fly-by-Wire flight control system is the logical solution to the major problems posed in "The Why"!

THE WHEN

Unfortunately, the transposition of quick verbal solutions into reliable hardware takes years of hard work, frustration, and money. The F-4 Survivable Flight Control System is setting out to demonstrate the survivability and motion control which have been raised from words to hardware.

We've flown our Fly-by-Wire F-4 more than a dozen flights as of this writing. The specific system we have installed is designed to develop the concept, not as retrofit or production. We have some additional problems which must be solved before Fly-by-Wire becomes part of any "kick-the-tire, light the fire" operational capability. Two problems might be:

Pilot workload for control of a complex electronic system must be reduced so that the system requires no more pilot attention than present control systems.

Control of an unstable aerodynamic vehicle must be regained after departure from controlled flight. Imposing verbal or written constraints on the pilot or severely limiting his maneuvering envelope is not sufficient. He must either be prevented from going out of control or be able to regain control.

In summary, I see the concept of **Fly-by-Wire married to the concept of the control-configured vehicle. Specifically:**

A state-of-the-art reduction in vehicle structural weight suggests an aerodynamically unstable aircraft.

An unstable aircraft requires a full-authority motion command with feedback flight control system. It makes no sense to give the pilot a mechanical flight control back-up since he can't cope with the response of the aircraft anyway. Therefore, the future of the manned fighter aircraft demands Fly-by-Wire!

Give it some thought at the next Happy Hour! Hang Loose!!!

Engineering the System

By ROBERT L. KISSLINGER / *Senior Project Guidance and Control Mechanics Engineer and*
GEORGE R. SMITH / *Project Electronics Engineer*

The Survivable Flight Control System (SFCS) uses a three-axis Fly-by-Wire primary flight control system. In each axis, four channels are provided for redundancy. A simplified block diagram of one axis of the system is shown on the next page.

In the Fly-by-Wire system, electrical sensors measure pilot commands, in the form of stick and pedal movements, and the aircraft flight path responses in pitch, roll, and yaw. Computers process the sum of the sensor inputs and act to provide electrical signals which move the primary control surfaces, such as ailerons, to produce the desired aircraft motion. In the event of a total computational failure or sensor malfunction, a direct electrical path, called the electrical back-up mode, is provided to allow the pilot to get home and land.

FBW SYSTEM MECHANIZATION

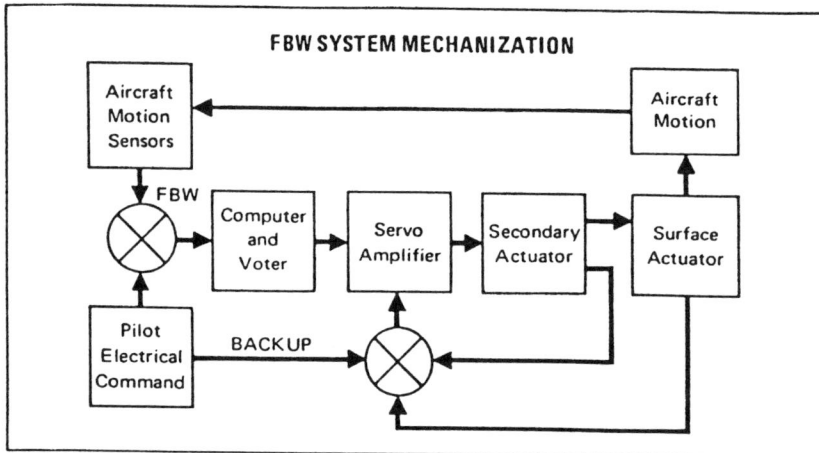

REDUNDANCY

Quadruplex (four-channel) redundancy is used in all on-line system components, including power supplies, to improve system reliability and increase the probability of mission completion. The system is designed to sustain two similar failures per axis without significant degradation in performance. Improved survivability obtained through the use of physical isolation or dispersion of components is utilized, insofar as practical for the test aircraft, to demonstrate the principle of survivability through dispersed redundancy.

BUILT-IN TESTS

An extensive ground built-in test (BIT) capability is included in the SFCS. The system automatically tests the SFCS and subsequently indicates a Go or No-Go condition to the pilot and ground crew. Most detected failures are automatically isolated to a line replaceable unit; and LRU failure indications are displayed on a maintenance test panel. The ground BIT requires approximately four minutes and is positively locked out during flight.

MONITORING

Inflight monitoring (IFM) is employed to automatically detect and disengage failed channels or secondary actuator elements. Reset switches with integral status indicator lights are installed on the main instrument panel to provide continuous system status information to the air crew. Momentary or inadvertent disconnects can be reset using these switches. The placement of

the reset switches on the master control and display panel in the front cockpit is illustrated below.

Considerable flexibility is provided in the system design to enable comparisons of various control modes, methods of applying pilot inputs, and airplane response characteristics. Both center stick and side stick controllers are provided in the front cockpit for side-by-side evaluation.

CHANNEL COLOR-CODING

A color-coding system is utilized throughout all SFCS LRU's, interconnect wiring, connectors, and hydraulic plumbing fittings to help prevent installation and maintenance errors. The four channels of each axis are identified by the colors *red, blue, black,* and *yellow.* Each wire bundle or hydraulic

line which must be connected to an SFCS LRU is identified by one of these colors. LRU's which contain elements of more than one channel have the appropriate color-code identifications adjacent to electrical connectors and hydraulic line fittings.

Electrical and hydraulic power sources to be used in each channel were selected to minimize the effect of power source failures.

ELECTRICAL POWER

The primary sources of electrical power in the test airplane are two engine-driven ac generators. These generators power the left-hand and right-hand ac buses in a split-bus configuration. In the event of power failure of one of the sources, the bus-tie contactor automatically connects the buses together so they are both powered from the remaining source. To obtain quadruplex power sources for the SFCS equipment, two transformer-rectifiers are connected to each of the two electrical buses. Each of the transformer-rectifiers is connected in parallel with an aircraft battery and connected to *one and only one* SFCS channel. The batteries have sufficient capacity to power the SFCS for approximately one hour. Use of the batteries assures continued operation of all four SFCS channels in the event of total ac power failure.

HYDRAULIC POWER

Three hydraulic pressure sources are normally available in the F-4 airplane.

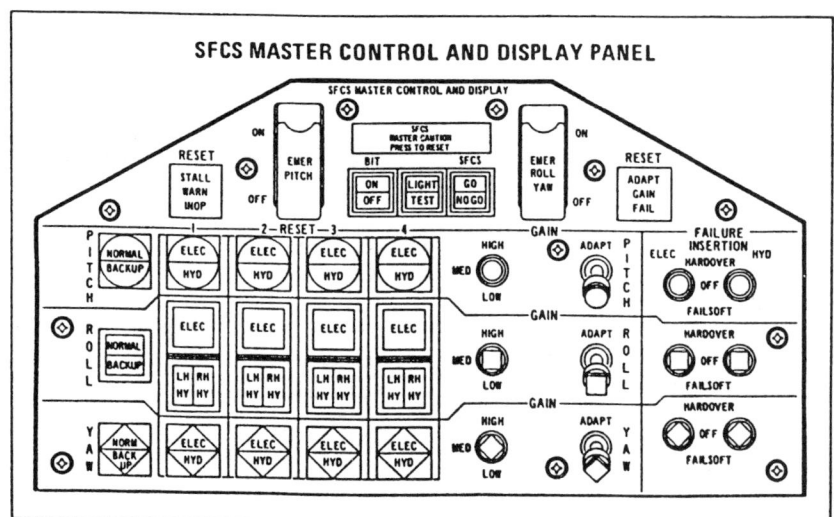

These pressure sources and the assigned color codes are: Power Control Hydraulic System Number 1 (PC-1) — red; Power Control Hydraulic System Number 2 (PC-2) — blue; and Utility Hydraulic System — black. A fourth hydraulic system is required to maintain quadruplex redundancy for the SFCS in the test airplane. An auxiliary power unit containing an electric motor-driven hydraulic pump is utilized to provide the fourth hydraulic system, which is color coded yellow. Excitation for the auxiliary power unit is normally supplied from the left-hand ac bus with automatic switchover to the right-hand ac bus in the event of electrical failure.

SECONDARY ACTUATORS

Secondary actuators are initially used in all axes of control to convert electrical command signals to mechanical position commands for application to the surface actuators. Separate quadruplex electrohydraulic secondary actuators were installed, in preference to integrated secondary and surface actuator units, to permit utilization of existing F-4 surface actuators.

The four elements of the secondary actuators are physically isolated from each other, powered from separate hydraulic sources, and commanded through separate electrical channels of the Survivable Flight Control Electronic Set (SFCES). The outputs of the four elements of each secondary actuator are physically connected to provide the single mechanical input required for the surface actuator. The linkage between each secondary actuator and its associated surface actuator is designed with sufficient strength and integrity to maintain overall SFCS reliability.

CONTROL AXES

A brief functional description of each primary axis of control is presented in the following paragraphs to indicate the development approach used in the SFCS design.

LONGITUDINAL AXIS — A high-gain control loop with both pitch rate and normal acceleration inputs is used in the longitudinal axis. Both pilot-selectable gain control and adaptive gain control are provided. A structural filter is utilized to reduce the loop gain at airframe structural resonance frequencies and help eliminate the effects of structural bending.

Both neutral speed stability (NSS) and takeoff and landing (TOL) functions are included to eliminate a trim requirement in the clean configuration and to provide a positive speed stability characteristic for takeoff and landing. With the NSS function selected, no steady-state pilot-applied stick force or trim input is required to compensate for the change in stabilator position needed to trim the airplane due to changes in airspeed and/or altitude. Selection of the TOL mode results in a requirement to apply stick force or manually trim the aircraft as airspeed and/or altitude are varied. In TOL, airspeed changes at subsonic flight conditions will result in positive speed stability; that is, push forces will be required as airspeed is increased whereas pull forces will be required as airspeed is reduced.

Stall warning is provided through increased stick force and audio tones as the stall is approached.

LATERAL AXIS — In the lateral axis, a fixed-gain roll rate inner loop is utilized to provide a highly responsive roll rate command system. Pilot inputs are shaped to provide desirable roll rate time constants for all flight conditions. A structural filter is included to attenuate loop gain at structural resonance frequencies. An input from the stall warning circuit is utilized to remove roll rate feedback at high angles of attack and thus prevent normal aircraft wing rock from causing spin-inducing lateral control deflections. A shaped lateral stick-force output is provided to the yaw axis for use in turn coordination.

YAW AXIS — Yaw rate and lateral acceleration signals are used in the yaw axis to improve Dutch Roll damping and frequency characteristics. Both pilot-selectable gain control and adaptive gain control are provided. Pilot rudder pedal inputs are applied through a shaping filter to further improve response characteristics. Shaped lateral stick-force inputs are applied through variable gain networks to improve turn coordination. The roll-to-yaw crossfeed circuit is utilized at all flight conditions except high q, where the adaptive gain is set to zero since crossfeed is not required.

Maintaining the System

By MAINTENANCE ENGINEERING / *Support Technology*

Maintenance has been an important consideration in the Survivable Flight Control System (SFCS) development program due to the scope of the modification. The SFCS affects every major subsystem of the test aircraft except the propulsion system, while adding 37 major line replaceable units (LRU's), removing 16 LRU's, and relocating two LRU's. The Fly-by-Wire portion of the SFCS contains in excess of 24,000 electronic piece-parts distributed among 25 LRU's. The electrical wiring installation adds some 3,000 circuits consisting of approximately 4,500 wires and 9,000 terminations. An additional flight control hydraulic system, powered by an electrically driven hydraulic pump, is installed. All of this is incorporated in a much-modified prototype Phantom. The following paragraphs describe some of the maintenance features, support equipment, and documents used to ease the SFCS maintenance burden.

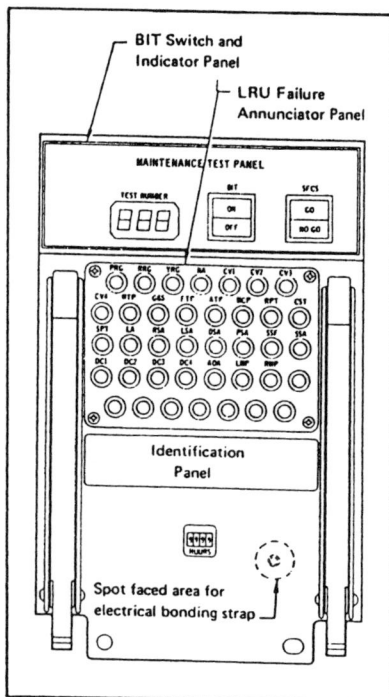

BIT Switch and Indicator Panel

LRU Failure Annunciator Panel

Identification Panel

Spot faced area for electrical bonding strap

MAINTENANCE FEATURES

The most complex piece of maintenance-oriented equipment integrated into the SFCS is the built-in test (BIT) system. The heart of this system is a permanently programmed digital computer housed in the Maintenance Test Panel (MTP) with peripheral components in several other LRU's. This system detects failures, while actively exercising and interrogating the system with a four-minute test program, and isolates most failures to the LRU level.

One of the primary purposes of the BIT system is to detect passive failures. Additionally, it checks logic circuits and tests the SFCS to tighter tolerances than the inflight monitor system. Because the BIT check is an active test, precautions have been taken to positively prevent its activation during flight. The illustration shows the BIT system display on the MTP.

Access, viewing, and mounting provisions for LRU's requiring frequent access, inspection, and/or removal are given special attention. The six SFCS computers are mounted in the nose of the aircraft in individual trays, each secured by two cam-action quick-release handles and fitted with rack-and-panel-type connectors. Cam-action handles

and rack-and-panel connectors were a necessity since the computer connectors contain up to 424 pins each. Direct access to the SFCS computers is by way of the former forward camera bay door. A small window for viewing the LRU failure indicators on the MTP is provided in this door since all the original photographic windows in the nose of the test aircraft have been replaced with metal.

A special inspection and servicing door is installed in the aft fuselage to permit inspection and servicing of the fourth hydraulic power supply. The four SFCS 20-cell nickel cadmium (NiCd) batteries, weighing 55 pounds each, are provided with swing-down type mounting racks to speed servicing.

To preserve the redundancy characteristics of the SFCS, an optimum interconnection arrangement of the ac and dc electrical power supplies and hydraulic power supplies is used to prevent single generator or hydraulic pump failures from causing multiple channel failures. This leads to coding all wiring, hydraulic lines, and connectors in each channel with a color unique to that channel. In addition, electrical connectors incorporate keying, clocking, and sizing variations to further prevent inadvertent interconnection of channels.

Equipment packaging is designed to simplify maintenance and, if required, modifications. The SFCS computers are constructed with plug-in circuit cards mounted to a chassis. Circuit cards are the metal type with pin-to-pin wiring which allows circuit changes to be readily made. Circuit components identified as likely prospects for redesign are mounted directly to the boards to allow quick replacement.

Channelization and rip-stop design provide a fallout benefit to maintenance of the hydromechanical secondary actuators. The actuators are composed of four separate channel modules mounted to a single frame, thereby eliminating the necessity for disassembly of the total actuator for repair of a single module. In addition, servo and solenoid valves are directly accessible and removable from the exterior of the actuator.

Some of you who are familiar with

NiCd batteries may have noted the reference to a 20-cell battery. The SFCS batteries are actually standard MS24497-5, 19-cell NiCd batteries modified to a 20-cell configuration. This eliminates some potential operating problems and aids maintenance troops since the batteries are much less prone to boiling and other problems associated with overcharging.

SUPPORT EQUIPMENT

The biggest piece of specialized SFCS support equipment is the Mobile Ground Test Facility (MGTF). The MGTF is an 8 by 24 by 13 foot trailer-mounted electronics shop for the SFCS. Included in the MGTF are a complete hot mock-up type test bench for the electronics set, a complete set of electronic LRU test fixtures, an eight-track strip-chart recorder, and an analog computer. The analog computer may be plugged into the electronics set in the aircraft or on the test bench and is used to control closed-loop testing of the SFCS. The MGTF is shown at the right.

A special piece of handling equipment is provided to remove and carry the batteries. This is a handle which attaches to the top of the battery, allowing it to be lifted and moved about in the same manner as your car battery would be carried at the local service station.

MAINTENANCE DOCUMENTATION

In addition to the usual drawings and modification data generated to install the SFCS in the test aircraft, special documentation is used to maintain the SFCS. Fifteen books of supplier-written Operating and Maintenance Instructions are provided to care for the various components of the SFCS. Another booklet contains maintenance instructions for on-aircraft maintenance. Regular F-4 scheduled maintenance procedures are completely rewritten to incorporate the SFCS requirements.

THE FUTURE

Don't get the impression from our discussion of an R&D prototype that Fly-by-Wire and SFCS will increase flight control system maintenance problems when introduced in production aircraft. On the contrary, prospects for

reduced maintenance compared to current mechanical flight control systems are very good when Fly-by-Wire is introduced in future aircraft. Control cables, bellcranks, push rods, and other similar mechanical devices will be a thing of the past with perhaps the only mechanical device left being the pilot's controls and the control surface actuators and linkages. The use of state-of-the-art electronics will keep maintenance to a minimum with the likelihood that Fly-by-Wire systems on future aircraft will encounter considerably less on-aircraft and shop maintenance than the current generation of stability augmentation, aileron rudder interconnect, and autopilot system equipment.

SFCS EQUIPMENT AND LOCATIONS

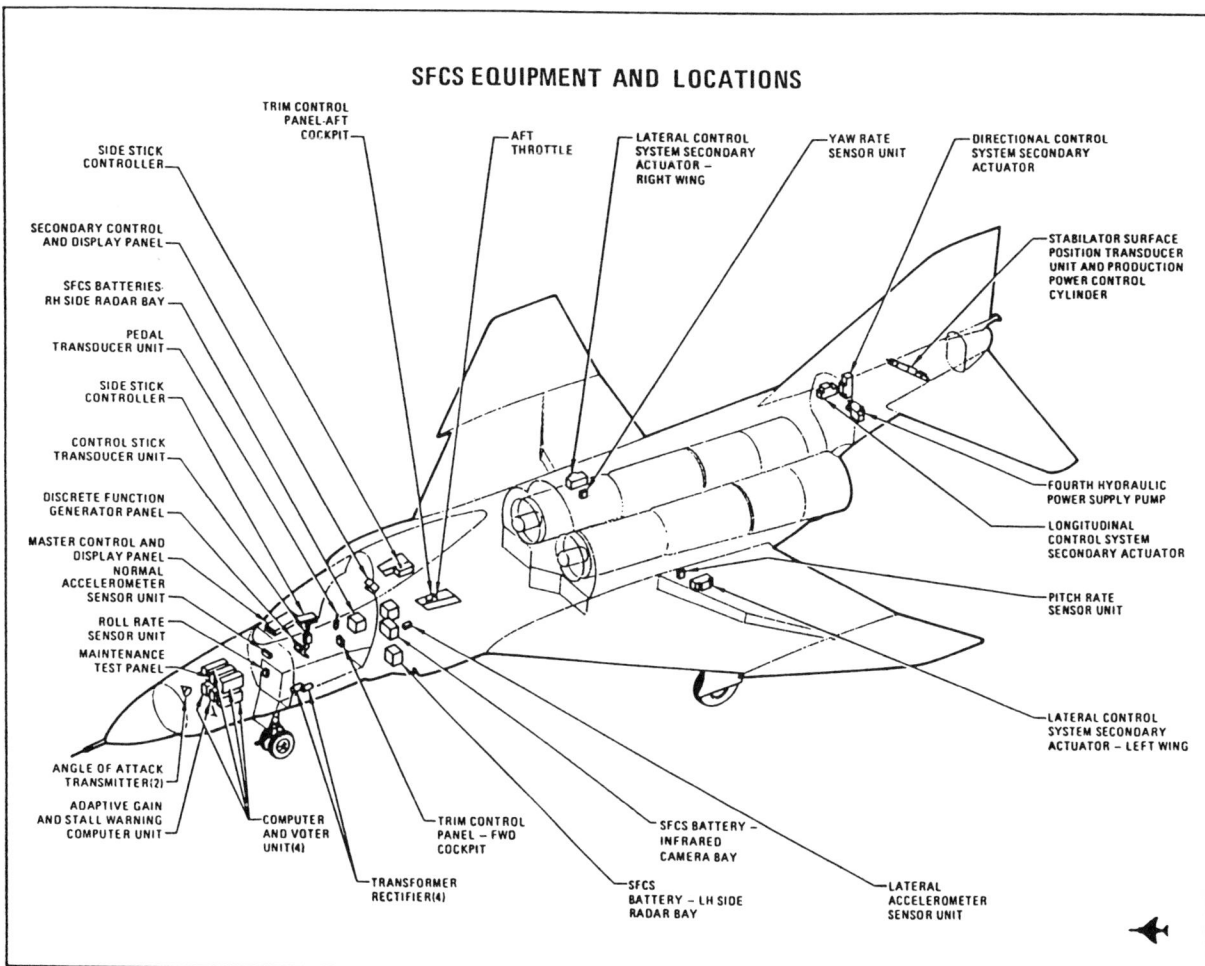

A Phantom with Famous Firsts...
62-12200

Contrary to appearances, all airplanes on this page are the same one. The unchanged tail number reveals a truly multi-service aircraft in this famous Phantom.

Originally scheduled as a Navy F-4B, and 266th in the production line, 62-12200 became the prototype Air Force YRF-4C.

With tests complete as a YRF, 12200 became the YF-4E. Instead of a camera, the chin now had a rapid-fire cannon. Again, this aircraft became the first of a long line of gunfighters.

Next came installation of leading edge wing slats for tests of maneuverability and buffet during Project Agile Eagle. Again, the successful test results are being seen on flight ramps as slats-equipped aircraft roll off our production line.

Now the camouflage is gone, replaced with blue and white (and a smattering of red). Today's task is to look at the systems of tomorrow in the tests of the fly-by-wire system described in the preceding article.

Today's RF crews man aircraft which are direct offsprings of 62-12200. Likewise the crews of F-4E's. Those whose aircraft are more maneuverable because of leading edge wing slats are benefiting from lessons learned with 12200. For all who will be flying the fighter of the future, the fly-by-wire system in today's 62-12200 is a portent and promise of better control, greater maneuverability, and a greater assurance of survivability.

How much more could any one aircraft serve?

Appendix C

Fly-by-Wire 777
Keeps Traditional Cockpit

FLY-BY-WIRE *777* KEEPS TRADITIONAL COCKPIT

DAVID HUGHES/BOEING FIELD, SEATTLE

The 777 cockpit is a cross between the 747-400's and 767's, but the similarities are only skin deep. The new flight control system depends on wires rather than cables

Boeing has combined a fly-by-wire system with traditional cockpit controls in the *777* to create a large, twin-engine transport that represents an evolution rather than a revolution in its product line.

Pilots of cable-controlled, glass-cockpit aircraft will feel right at home in the *777* cockpit. Its front panel is very similar to the one in a Boeing 747-400. The overhead panel and pedestal copy the 767 layout. There is a yoke that moves when the autopilot is engaged, and the throttles move when the autothrottle system adjusts power. This tactile feedback is an additional cue to the pilot as to what the auto-flight systems are doing. (Airbus has taken a different approach on fly-by-wire cockpits by employing side-stick controllers and throttles that remain fixed when the autothrottle adjusts power.)

Boeing also has added speed stability to the fly-by-wire logic so that forces build up on the control column when the aircraft changes speed. The pilot then neutralizes the forces with trim.

The internal structure of the *777* avionics system represents a dramatic departure from previous Boeing designs. Fly-by-wire computers warn pilots when they are flying too fast or slow, or when they are overbanking. Boeing does not limit the pilot's ability to override the fly-by-wire computer and command a high-g pull-up or to roll more than 60 deg. to recover from an unusual attitude. The Honeywell Airplane Information Management System integrates all avionics functions in two cabinets.

Boeing also has designed the *777* to perform extended-range twin-engine operation (ETOPS) with redundancy built into the electrical system. Simplified controls and displays will make it easy for pilots to divert to an alternate airport if necessary.

The aircraft also handles well on one engine with no unusual control problems, even though up to 77,200 lb. of thrust is provided on one side of the aircraft. The *777* is quite agile for a heavy transport, providing rapid roll rates of up to 20 deg./sec. for evasive maneuvering.

This *AVIATION WEEK & SPACE TECHNOLOGY* pilot had the opportunity to fly a *777* at Boeing Field last month. The flight test aircraft, sequence number WA001, made the first flight on June 12, 1994 (*AW&ST* June 20, 1994, p. 20). The aircraft has flight test monitoring equipment in the cabin. For our flight, the Pratt & Whitney turbofans provided 77,200 lb. of thrust.

> *Fly-by-wire computers warn pilots when they are flying too fast or slow, or when they are overbanking*

John E. Cashman, chief pilot on the *777* program, occupied the right seat while I took the left seat. After we entered the cockpit, Cashman started the APU and switched power on the *777* electrical system, which made a smooth "no break" transition from ground cart power. The overhead scan of system status is very straightforward.

The aircraft's gross weight was 445,000 lb., including 135,500 lb. of fuel. Cashman set up our takeoff speeds in the flight management system (FMS) control display unit. V_1 would be 125 kt., rotate (V_R) would be 129 kt. and takeoff safety speed (V_2) would be 137 kt. The speeds are all marked on the vertical tape on the primary flight display. We would climb out initially at about 152 kt. (V_2 + 15) as commanded by the flight director when VNAV engages at 400 ft. above the ground.

The *777* has two 8 X 8-in. flat-panel, liquid crystal displays (LCDs) in front of each pilot—one for primary flight information and one for navigation. There is also one multifunction display (MFD) in the center of the panel and a second one below it in front of the throttle quadrant on the center pedestal. The large-format displays from Honeywell make even minute details easy to read on the navigation map display. The three small Rockwell Collins LCDs for standby attitude, speed and altitude are remarkably clear and easy to read.

When *777*s enter service, they will also have data-link capability that will allow pilots to receive weather and company information for display on the MFD or to be printed out on a printer at the rear of the center pedestal. The data-link system also is designed to accommodate future ATC transmissions on speed, heading and altitude assignments.

A flight control synoptic page is new for Boeing aircraft, and it shows the movement of all of the primary flight controls—ailerons, flaperons, spoilers, elevators and the rudder. When control is lost over some surfaces due to a hydraulic system failure, that surface is simply crossed out on the display, making it easy for the pilot to see what he has left.

Cashman set up the navigation displays for a departure on Runway 31L at Boeing Field and a flight to the east to Grant County Airport at Moses Lake, Wash., for touch-and-goes. We planned to use full thrust on takeoff and climb with flaps at 20 with a center of gravity of 28% mean aerodynamic chord.

We used a paper checklist on our flight, since WA001 is not equipped with an electronic checklist. Earlier, I had the chance to see the electronic checklist work in both normal and emergency situations in the simulator. There are closed-loop features in the checklist that show an item in green when it has already been accomplished.

For example, the system automatically checks flap handle position, flap position and the flight management computer-designated flap setting to see if they agree prior to takeoff. Some checklists involving abnormal situations have deferred items that are automatically appended to the appropriate checklist and presented to the pilot later. This is a nice electronic reminder.

WE SET THE 777'S AUTO-BRAKE system on the RTO position, which means that any reduction of thrust to idle above 80 kt. on takeoff roll would result in a maximum brake application with the full 3,000 psi. available.

It is possible to start both of the Pratt & Whitney engines simultaneously, and Cashman placed both start/ignition switches and both fuel control switches to run. Airlines may elect to use this procedure. However, GE engines must be started individually, and Rolls-Royce engines have not been tested yet.

With a cold engine, the fuel comes on at about 21% of N_2 and starter cutout occurs at about 41% of N_2. Cashman said the auto-start system monitors EGT during the start sequence. If the first start attempt does not work, the automatic system will try a second and third time using

---

In C* the fly-by-wire computer delivers the pitch rate or 'g' commanded by the pilot

---

both igniters. It takes a long while for the big turbofans to spin up to idle, and these engines each took about 50 sec. N_2 peaked about 57% on both engines.

When we were cleared to taxi, I advanced the throttles. The 777 does not need a lot of thrust to break away, and after we began to roll, I pulled the throttles back to idle. There was plenty of power to carry us through a 180-deg. turn away from the hangar in front of us. The nose gear is 12 ft. behind the pilot's seat, so you have to delay the turn a bit if you want the wheels to track on the yellow line. The 777 has the same wing span as a 747, and the size felt comfortable to me based on my experience as a USAF Reserve C-5A copilot.

I turned down the taxiway and was initially a few feet left of centerline until I got properly oriented. The carbon brakes did not bind. Each main gear has six wheels, and when the pilot brakes below 30 kt. of ground speed, only four of the six brakes are used. The selection of brakes is automatically alternated, and I noticed no change in the braking action. We were

taxiing at about 20 kt. based on the ground speed readout on the LCD display in front of me.

As we taxied, I performed a flight control check. I did not feel much movement through the aircraft body when I deflected the rudder stop-to-stop, even though it weighs 1,500 lb. We completed our checklists and received clearance for takeoff on Runway 31L. I maneuvered the 777 onto the runway for a rolling takeoff. Winds were 135 deg. at 6 kt. (a slight tail wind) with a 4,400-ft. overcast ceiling.

I stood the throttles up to 1.05 EPR and let the thrust stabilize. I then depressed one of the takeoff/go-around (TOGA) buttons on the throttles and the autothrottle system advanced them to 1.364 EPR. The aircraft weighed 444,300 lb. as it rolled down the runway. A computer voice called out V_1, and Cashman said "rotate" as I pulled back on the control column and the aircraft rotated smoothly. We retracted the gear, and I picked up the flight director guidance, which called for a gradual pitch change to about 17 deg. Our clearance was to climb to 2,000 ft. on runway heading, expecting radar vectors after that.

CASHMAN EXPLAINED earlier that the fly-by-wire system technically is in the direct mode (column to elevator) during takeoff. It then transitions to the C Star (C*) maneuver demand control law. In C* the fly-by-wire computer will deliver the pitch rate or "g" commanded by the pilot when he moves the control column. This is the case regardless of what else is happening as long as speed remains constant. Thus the computer may command elevator movement to compensate for wind, turbulence or a change in thrust without moving the control column. However, the

777 roll control surfaces are directed by wheel movement, rather than on a maneuver-demand basis.

The autothrottle system reduced power to a climb setting at 1,000 ft., and I heard a change in engine noise as the fan resonated. Cashman said the loudest engine noise seems to occur at this point, but noted that it is amplified a bit in the test airplane, which has no sidewall insulation. We retracted the flaps to 5 at 1,500 ft. and to a setting of 1 (slats only) at 2,000 ft. at about 180 kt. Passing 2,400 ft. at a speed of 200 kt., we brought the slats up and began accelerating to 250 kt. while climbing to 9,000 ft. The speed tape reminds the pilot of gear and flap placard speeds by marking the next restriction that applies with a red symbol.

WHEN WE WERE CLEARED by ATC above 10,000 ft., I followed the command bar to accelerate to the 320-kt. economy climb speed. This speed, usually 300-320 kt., is calculated by the flight management computer based on the aircraft gross weight and the cost index entered by the airline. Cashman had previously entered a cost index of 110, which is a typical value. Normal climb speed is close to V_{MO} (330 kt.). As we accelerated, I had to trim to adjust to the new speed.

When the pilot trims the aircraft, he is essentially changing the fly-by-wire computer reference speed, not moving the trimmable horizontal stabilizer. The computer controls the stab trim on its own. When the trim reference speed in the computer equals the speed of the aircraft, the forces on the pilot's control wheel are nulled. The control column forces buildup at about 3 lb. for every 10 kt. of airspeed change, so you can feel them if the aircraft is accelerating or decelerating and you are not trimming.

CRAIG S. PETERSON, a 777 flight deck engineer responsible for fly-by-wire systems, said speed stability is one of the main differences between Boeing and Airbus approaches to fly-by-wire controls. Boeing

is keeping the "feel" and control of the 777 similar to the rest of its fleet and giving pilots additional speed cues.

We accelerated slowly on our climbout, so it was easy to keep up with the trim. Our climb rate was about 2,000 ft./min. passing 12,500 ft. with a fuel flow of 16,900 lb./hr. on each engine.

Cashman pointed out that the navigation map display shows the situation behind the airplane as well as in front of it. The Honeywell FMS also keeps track of the four closest alternate airports in the navigation data base. It is not even necessary to call up the "legs" page on the FMS to go to one of the alternates. The pilot simply hits the "alternate," the "direct" and the "execute" buttons on the control display unit to begin a diversion. Data link will allow the company to let the pilot know when an alternate is below minimums and substitute another one with suitable weather. The FMS also provides an engine-out driftdown schedule.

All of this could be very useful to the crew of an ETOPS aircraft in an emergency far from land. The aircraft also has a highly redundant electrical system with backup generators on each engine to meet ETOPS requirements. All aircraft systems, including fuel, have been kept simple to operate.

We continued our climb while turning toward the North Training Area east of the Cascade Mountains. We planned to maneuver in that area between FL330 and FL370. Passing 20,000 ft., we were flying at 314 kt. (Mach 0.68), climbing at 1,500 ft./min. and burning about 14,000 lb./hr. on each engine.

We passed FL330 holding Mach 0.823, which we held to level off at FL350. At about 1,000 below our level-off altitude, we were burning about 11,000 lb. of fuel per hr. on each engine.

Cashman said the climb speed is very close to the initial cruise speed, so no trim adjustments are needed at the top of climb. We reached FL350 21 min. after brake

release, and we then accelerated to Mach 0.84. The wing was originally designed for a long-range cruise speed of Mach 0.83, but it has turned out to be more like Mach 0.84, according to Cashman.

Once stabilized at Mach 0.84, I noted that the indicated airspeed was 288 kt., true airspeed was 480 kt., EPR was 1.104, N_1 was 80.4%, N_2 was 81.9%, EGT was 367C and fuel flow was 7,100 lb./hr. on each engine. After about 2 min. with the autothrottle system engaged, I checked these figures again—the readings were still very close. Our fuel flow was 7,000 lb./hr./engine and the airspeed/Mach was the same. Our gross weight was approximately 431,300 lb. at this point. Cashman said the autothrottle system will let the speed wander a bit to minimize throttle activity during cruise, and that is typical of modern transports.

THE 777 FLY-BY-WIRE system uses three different modes for operation. Normal mode provides a variety of augmentation such as stall and bank-angle protection. A loss of air data and other malfunctions would result in the system reverting to the secondary mode. Most augmentation is lost along with the autopilot. Then there is the direct mode, which is the most degraded form of operation. The system would only go into this mode following serious malfunctions—which are highly improbable. The pilot also can select direct mode at any time using a guarded switch on the overhead panel, although no normal or emergency procedure calls for this.

None of the overspeed, stall and bank-angle protection features are available in direct. Cashman selected the direct mode with the overhead switch so I could maneuver the aircraft briefly in this configuration. The 777 flies very much like a manual, cable-controlled airplane in this mode. There is no yaw damper, however, and the aircraft has a tendency to Dutch roll a bit, as I found when I moved the rudders. But other than that, the direct mode offers docile handling qualities for such a degraded mode. Few pilots will ever have to use direct mode in revenue service.

Cashman reselected normal mode, and I executed a 60-deg.-bank turn to the left. To maintain level flight, the pilot has to keep the wheel turned against the resistance provided by bank-angle protection and pull back on the column. But the pilot can still select any bank angle desired. I let go of the controls, and the fly-by-wire system quickly returned the aircraft to less than 30 deg. of bank. I then pulled back far enough on the control column during the turn to see the pitch limit indicator come into view on the attitude indicator. This symbol shows the pitch attitude at which stick shaker would activate if I continued to pull. It is displayed at slow speeds when the flaps are up and at all times when the flaps are down.

I got out of the seat, and another guest pilot did some maneuvering at altitude, then descended to 16,000 ft. for some low-speed evaluations. At this altitude, we did a "slam" acceleration on the left engine from idle. It took about 9-10 sec. for the engine to reach maximum continuous thrust. The aircraft does not have much of a tendency to pitch up when full power is applied in the normal mode because the fly-by-wire system compensates for thrust coupling associated with the large engines underneath the wings.

We began deploying flaps at 16,000 ft. after I returned to the left seat. Flaps 1 (slats only) were deployed at 215 kt. Flaps 5, which would be deployed on base leg during an approach, were selected at 195 kt. Flaps 20, which are normally deployed at about glideslope capture when the gear is extended, were selected at 178 kt. Few column inputs were needed during this process. The fly-by-wire system compensates with 2-3 deg. of nose-down pitch to maintain the same flight path as the flaps

deploy. No trimming is required during extension and retraction unless the speed is changed. The same compensation occurs when speed brakes are deployed.

BUFFETING WAS NOTICEABLE at Flaps 20. We set the speed bug over the numeral 20, which appeared on the speed tape at about 154 kt. This was the Flaps 20 maneuver speed, which means you can bank to 40 deg. and still not activate the stick shaker.

I throttled back to idle on the left engine and then began a turn in the direction of the "dead" engine. The fly-by-wire system assists the pilot in an engine-out situation by adding a little rudder when the control wheel is moved. Boeing is also developing a thrust asymmetry compensation (TAC) system that will take care of almost all of the yaw associated with an engine-out. This system will be flight tested this summer on the aircraft I flew. I worked on coordinating the turn by keeping the slip/skid indicator centered under the bank pointer on the electronic ADI. The asymmetry was not difficult to control.

I rolled out of the turn and throttled back the right engine a bit to slow to 133 kt. The fly-by-wire system would not let me trim the aircraft to a speed slower than this minimum maneuvering speed. Stall protection limits the trim reference speed at this point so that trim is inhibited in the nose-up direction. We were flying about 25 kt. above minimum control speed, which is below the 1g stall speed. At this slow speed, large right pedal force was required to keep the aircraft from yawing. The dual-rate rudder trim switch can be used in the high-rate mode (2 deg./sec.) to quickly feed in rudder trim and compensate for the asymmetric thrust.

For the next maneuver, I advanced the left throttle to match the right one as Cashman hit the rudder trim cancel switch. This slowly removes rudder trim. We began

a simulated go-around maneuver at Flaps 20 by advancing the throttles to TOGA. As I began a climb, Cashman cut the left engine to idle. The thrust decays gradually as the big fan winds down, and it is not difficult to keep up with the need for right rudder to maintain coordinated flight. Cashman said the engine even spools down slowly following a fuel cutoff. Boeing test pilots thought the loss of thrust would be a pretty dramatic event in the 777, but the inertia of the big fan makes it quite tame compared with other transport aircraft, according to Cashman.

WITH SYMMETRIC POWER again and both engines in idle, I began slowing to approach a stall with Flaps 20. We planned to go well beyond stick shaker and just short of a full stall. We started at 138 kt. at 16,000 ft. and began slowing at about 1 kt./sec. As we slowed below the top of the amber band on the speed tape (133 kt.), the nose-up trim stopped working as a stall protection. The auto slats also fired, meaning the leading-edge devices extended from sealed to gapped position to provide better handling qualities.

The stick shaker activated at 116 kt. We began to encounter buffet as I continued to pull to near full-aft column, and we came very near a 1g stall. Cashman estimated I was pulling about 40-50 lb. of pressure when I released the column and advanced the power to recover just before actually stalling. If I had pulled the column all the way to the stop, it would have taken about 70 lb. of force.

Cashman said the 777 stall characteristics at Flaps 30 were originally "quite sporty," and flight test aircraft rolled as far as 110 deg. during tests in this configuration. Boeing engineers then tailored the fly-by-wire program so the outboard ailerons did not come down at high angle of attack and Flaps 30. The inboard leading-edge slats were also blocked a bit to produce less lift inboard. If the aircraft stalls during a turn, the fly-by-wire system commands a roll to wings-level.

I took another try at flying in the direct mode and found that the roll rate was slightly less than in normal mode at Flaps 20. Cashman said a stall in the direct mode would not be any different except that the force buildup on the control column would be quite a bit higher. Direct mode also lacks any compensation for thrust coupling, so the airplane also pitches up more when throttles are advanced to full power.

It was time to call Seattle Center for clearance to Grant

BOEING 777 SPECIFICATIONS *

Powerplants	Pratt & Whitney PW4077, 77,200-lb.-thrust rating; General Electric GE90-76B, 76,400-lb.-thrust rating; and Rolls-Royce Trent 877, 76,900-lb.-thrust rating. Note: WA001 has flown tests with PW4077 at 77,200 lb. rating and other tests with PW4084 at 84,000 lb. rating for B Market aircraft. Rating is changed by a software modification to electronic engine control.
Weights	
Maximum takeoff weight	535,000 lb. (242,671 kg.)
Maximum landing weight	445,000 lb. (201,848 kg.)
Typical operating weight empty	302,000 lb. (136,984 kg.)
Maximum payload	118,000 lb. (53,524 kg.)
Maximum fuel capacity	207,700 lb. (31,000 gal.)
Dimensions	
Length	209.1 ft. (63.7 meters)
Fuselage Exterior Diameter	20.3 ft. (6.2 meters)
Tail Height	60.8 ft. (18.5 meters)
Wing Span	199.9 ft. (60.9 meters)
Wing Area	4,605 sq. ft. (427.8 sq. meters)
Wing Sweep	31.6 deg. at 1/4 chord
Fan Diameter	PW4077 - 112 in. (2.84 meters)
	GE90 - 123 in. (3.12 meters)
	Trent 877 - 110 in. (2.79 meters)
Performance	
Takeoff field requirement, sea level ISA +15 (typical mission)	5,500 ft. for 1,000-n.m. leg
Maximum range 375/400 passengers	PW4077 - 4,615 n.m./4,390 n.m.
	GE90-76B - 4,595 n.m./4,370 n.m.
	Trent 877 - 4,695 n.m./4,470 n.m.
Maximum Operating Speed	Mach 0.87
Long Range Cruise	Mach 0.84
Roll rate	Approximately 20 deg./sec.
Fuel jettison rate	5,200 lb./min.
Typical passenger loading (2 class)	375 (9 abreast); 400 (10 abreast)
Under floor capacity	32 LD3s, 600 cu. ft. bulk
	(total volume = 5,656 cu. ft.) or 10 pallets,
* For A Market 777	600 cu. ft. bulk (total volume = 4,750 cu. ft.)

County Airport to perform some touch-and-goes there. The 8 X 8-in. map display made it easy to picture our route to the airport. We began retracting the flaps as I initiated a descent to 9,000 ft. We planned an ILS approach to Runway 32R at Grant County, where the winds were 360 deg. at 12 kt. and the ceiling was 1,300 ft. broken.

CASHMAN SUGGESTED THAT I try the speed brakes, and I extended the spoilers using the handle on the center pedestal. The buffeting did not seem too pronounced with flaps up, but it gets much more noticeable at Flaps 30, as I found later. Speed brakes cause less buffeting than on the Boeing 757 and 767, according to Cashman. I retracted the spoilers as I began leveling off at 3,000 ft., and we began extending the slats (Flaps 1) at 213 kt. Flaps 5 were extended next at 194 kt., and I slowed to 175 kt. or V_{REF} for 30 deg. of flap plus 40 kt. With the glideslope alive, we extended the gear and selected Flaps 20.

I was hand-flying the approach with autothrottles engaged, a technique that Boeing recommends for manual flight in the 777. The moving throttles made it easy to monitor power changes. Flaps 30 were selected when we captured the glideslope, and we settled into a well-stabilized approach at about 137 kt. (V_{REF} + 5 kt.) at a weight of 419,000 lb.

As the aircraft crossed the threshold of Runway 32R, the altitude callouts at 50, 30 and 10 ft. helped me judge when to begin my flare. As Cashman suggested, I

waited until 20 ft. to begin a gradual flare. The result was a smooth touchdown. I lowered the nose, and Cashman reset the flaps and trim as we rolled down the runway. I advanced the throttles and, when the aircraft accelerated again, rotated the nose. We left the gear down and during climbout began retracting the flaps.

Cashman said later that the 777 is very easy to land, and this is partly due to ground effect. The airplane has a wing span as large as the 747, but the wing is closer to the ground on landing and ground effect is quite strong. Boeing engineers even put some ground-effect compensation (a slight pitch-down tendency) in the fly-by-wire control law for landing, but Cashman said it is hard to tell the difference flying with or without it.

The next approach was an autoland. We selected Flaps 5, slowed to 170 kt., engaged the autopilot and then captured the localizer. Soon we had intercepted the glideslope with gear down and Flaps 30, and the autopilot had us stabilized on a good approach path. Of course, the control column and throttles were moving to show the inputs being made by the autopilot and autothrottle systems. These movements were not very pronounced compared with my memory of how earlier generation autoland systems used to jockey the throttles around.

By slewing the magenta box on the speed tape so it was just above the reference (REF) marker on the tape, I set the autothrottle

to maintain V_{REF} + 5 kt. The spoilers were armed to deploy on landing, and the autopilot rounded out to a firm touchdown on centerline. I took over the controls as Cashman reset the trim and flaps. I then advanced the throttles and rotated the nose for takeoff. Cashman estimated that our rate of descent at touchdown was 3-3.5 ft./sec. New autoland software, which will reduce the descent rate by 1.5 ft./sec., should make touchdowns smoother.

Cashman pulled the right engine to idle just as I began to turn right to 120 deg. at 3,500 ft. with climb power set. I compensated with rudder pedal movement for a while before setting in rudder trim once we were stabilized on the downwind leg.

We planned a single-engine approach to a full stop at Flaps 20 with an approach speed of 143 kt. (V_{REF} + 5 kt.). We put the gear down and flaps to 20 at 170 kt. after intercepting the localizer and were soon established on the glideslope. The autothrottle controlled the left engine, and I could feel the single throttle move in my hand. Not much throttle movement was needed to keep the 777 on its approach speed. The 777 seemed easy to handle on an approach in a simulated engine-out condition at a gross weight of 413,000 lb.

RUDDER TRIM WAS ZEROED as I crossed the threshold, and the 50-, 30- and 10-ft. voice calls helped me time my flare. After another smooth touchdown, the spoilers deployed. Cashman suggested we go to maximum reverse on the left engine as a demonstration. We did not select thrust reverse immediately after touchdown, and it took a while for the engine to spin up. Directional control was not a problem. I applied the brakes at 70 kt., and we turned off onto a taxiway. In all, I had flown the aircraft for 1 hr. 45 min.

After the flight, Cashman said that Boeing 747-400 pilots will probably have the easiest time transitioning into the 777 (one foreign airline and one domestic carrier are planning to qualify some crews to fly both aircraft). Boeing 757/767 pilots also will find a lot that is familiar in the 777. ■

Appendix D

Related Programs and Data

A number of R&D programs either directly related to fly-by-wire or associated with the use of the technology were instrumental in advancing application to new designs.

As described by the MCAIR test pilot, fly-by-wire permitted moving the "lift point" to, or near, the center of gravity of the aircraft, not only changing (lowering) the pilot's workload but also allowing structural changes that reduced the vehicle's weight. Fly-by-wire, in conjunction with the leading edge slats (called relaxed static stability), was a fundamental design change used on the F-16.

Sperry Flight Systems of Phoenix, Arizona, was given a contract by the Air Force Flight Dynamics Laboratory on the application of fiber-optics to fly-by-wire. This work is described in AFFDL-TR-74-10 "Modification of Prototype Fly-by-Wire System to Investigate Fiber-Optic Multiplexed Signal Transmission Techniques."

A prototype, quad-redundant, fly-by-wire system developed for the Flight Dynamics Laboratory for use on their B-47 in-house program was modified to incorporate a fiber-optic data transmission link in one of the four channels. (See Figure A-1.) Data was transmitted between one of the elevator actuator channels and its control computer on multiplexed optical data busses in serial-bit, serial-word format at 500 KHz bit rate. The system was mechanized with a 100 ft fiber-optic cable and miniature optical transmitter and receiver modules, having a total bandwidth capability of about 4.0 MHz. System performance was verified, and the compatibility of the optical channel and the other three electrical channels was demonstrated.

It should be kept in mind that signal transmission by fiber-optics must be accomplished in digital form by pulses. (See Figure A-2.) Therefore, whenever a change is made, or required, a converter must be used, digital-analog or analog-digital. In all hydraulic actuation systems the servo control valve that controls hydraulic fluid is analog, hence in any system the final conversion must be to analog.

Honeywell Inc. had a program with the Flight Dynamics Laboratory titled "Military Transport (C-141) Fly-by-Wire Program," as reported in AFFDL-TR-74-52. This technical report covers work from April 1971 to January 1974.

In a number of respects, this was similar to the B-47 Program. As in the B-47, the fly-by-wire portion of the C-141 flight control system was installed as a parallel channel to the existing hydromechanical system (see Figure A-3).

The objective of the C-141 program was to evaluate and optimize the handling qualities of a large transport-type aircraft by using closed-loop fly-by-wire control laws and a side-stick controller. The feedback and closed-loop func-

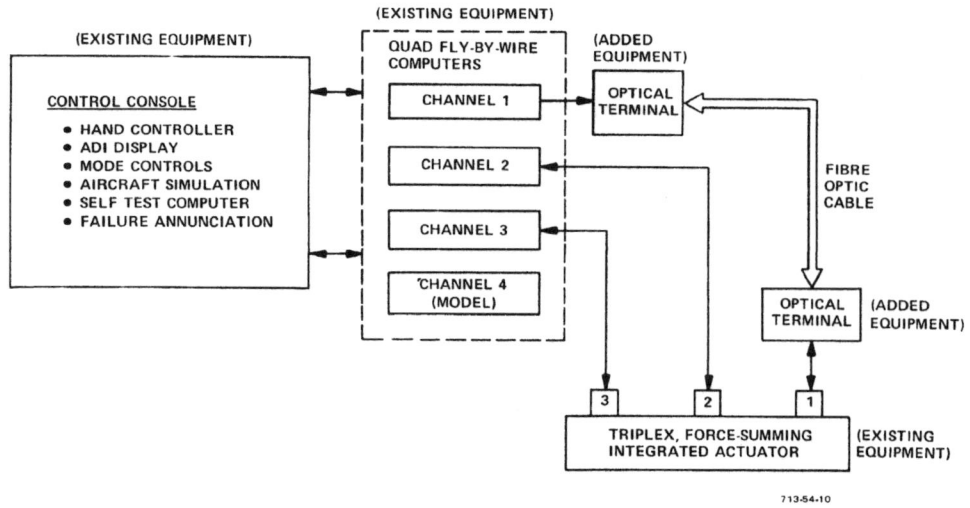

Fig. A1   Fly-by-wire system modification block diagram.

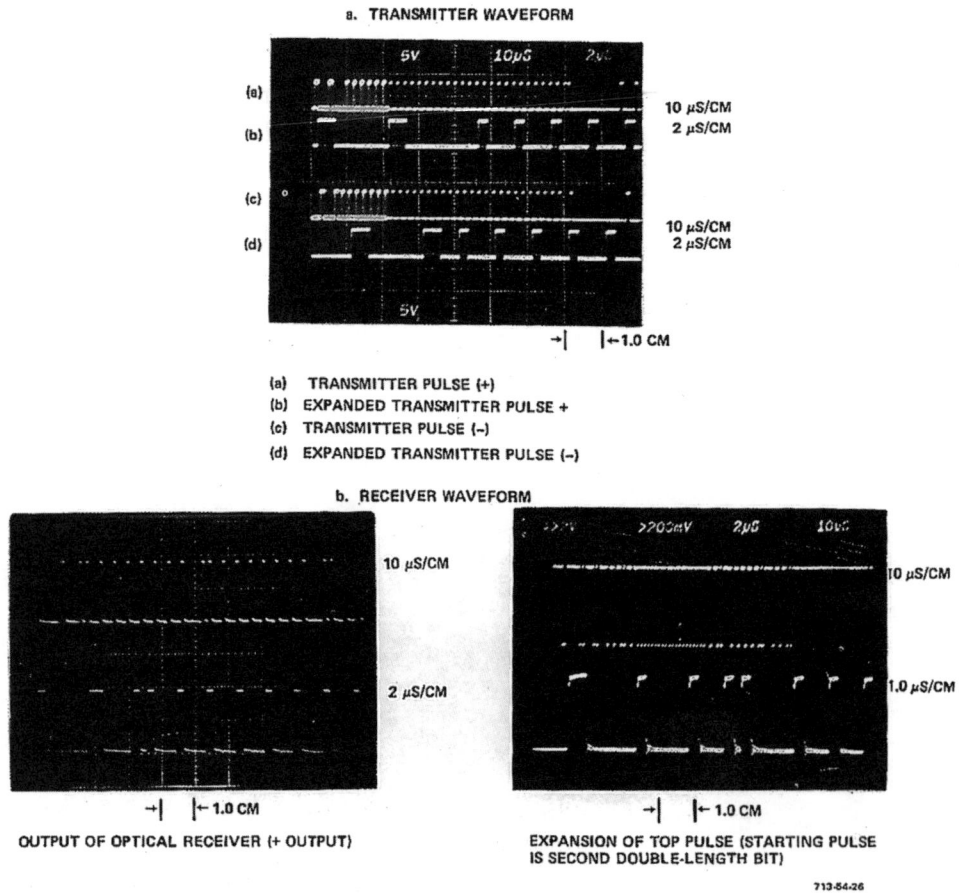

(a)   TRANSMITTER PULSE (+)
(b)   EXPANDED TRANSMITTER PULSE +
(c)   TRANSMITTER PULSE (−)
(d)   EXPANDED TRANSMITTER PULSE (−)

Fig. A2   Transmitter and receiver waveforms.

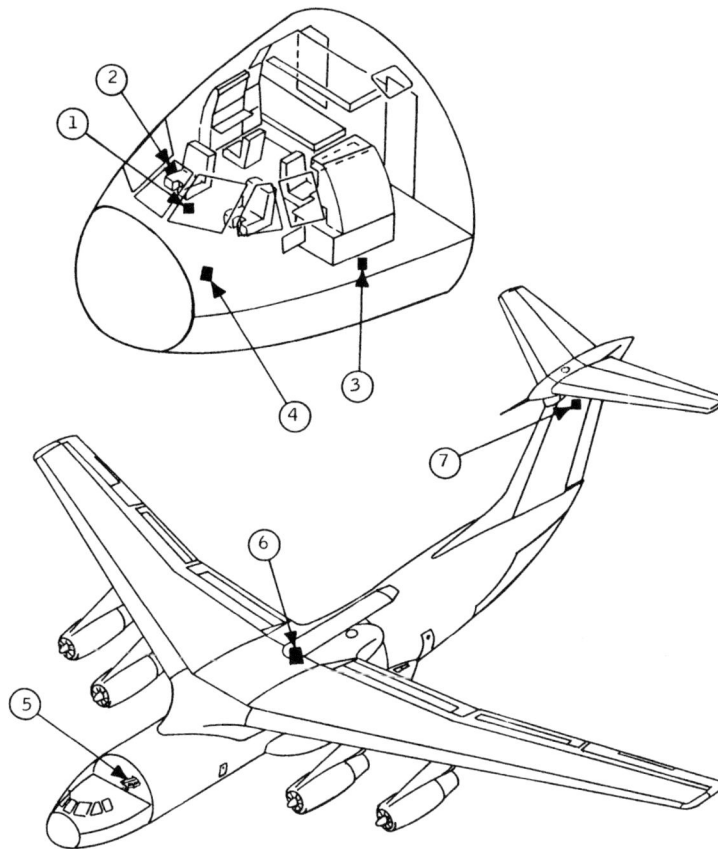

(1)   PILOT'S CONTROL PANEL
(2)   SIDE-STICK CONTROLLER
(3)   RATE GYROS
(4)   ACCELEROMETERS
(5)   TEST SYSTEM CONSOLE
(6)   AILERON CONTROL SERVOS
(7)   ELEVATOR CONTROL SERVOS

Fig. A3   C-141 fly-by-wire system installation final configuration.

tions of the fly-by-wire system improved vehicle stability and improved precision vehicle control for typical transport mission tasks such as formation para delivery, aerial refueling, low-level terrain following, and approach and landing.

The major emphasis of this program was placed on the definition of control laws to improve vehicle handling performance in the roll and pitch axes, rather than development of a highly redundant, electronic flight control system.

Evaluation of the C-141 fly-by-wire system included:

Control Laws   The fly-by-wire control laws were synthesized using existing military specification handling and control criteria, the C* longitudinal control criterion, and a piloted simulation.

Side-Stick Controller   The independent variables of a side-stick controller (feel, travel, position, and authority) were evaluated to establish side-stick controller and handling criteria.

Dual-Redundant Failsafe Mechanization This system used dual-redundant comparison-monitored inner loops, limited outer loop authority, $\pm g$ limit disengagement, limited forces servos, and shear pins to achieve flight safety.

The fly-by-wire system configuration initially proposed is shown in Figure A-4.

Simulation

Piloted simulation evaluations were conducted using a 6-degree-of-freedom aircraft simulation coupled to a multi-crew moving-base cockpit with its associated flight instruments, visual system, and variable flight control systems. Aircraft missions tasks were:

1. Approach and landing
2. Formation paradrop
3. Manual terrain following
4. Simulated aerial refueling

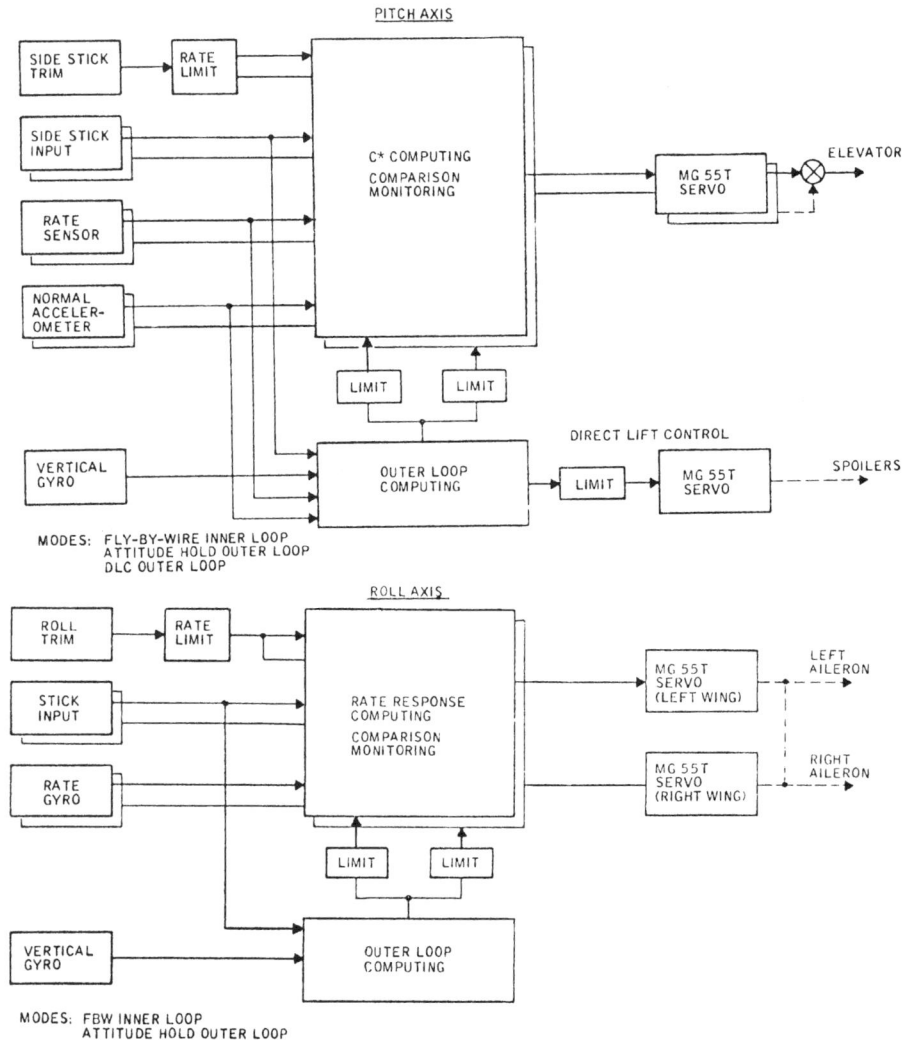

Fig. A4   Fly-by-wire system block diagram.

Simulation Results

Pilot ratings were found to be extremely useful in evaluating the two fly-by-wire system modes and various control system configurations (combinations of variable system parameters for the four simulated mission tasks). The C-141 simulations were compared with those on the F-4 by MCAIR.

C-141 Simulation

Single Failures—Nonredundant Portion of System. Single failures which could occur in nonredundant portions of the fly-by-wire system will not be electronically detected by the comparison monitor as no mistrack will exist; consequently, signal limiters have been mechanized to limit the authority of this type of failure.

The primary potential failures in this category are associated with outer loop devices and circuitry and the horizon stabilizer trim circuitry.

The effects of the above failures can in general be described by two types of vehicle responses:

1. Vehicle will not perform the desired function (i.e., will hold attitude or respond to side-stick vernier commands.)
2. Vehicle will perform an undesired maneuver (i.e., abrupt nose-up or wing-over maneuver).

The pilot will be able to detect the first type of failure by simply observing vehicle performance. Vehicle transients will not be excited under those conditions because stability will be retained through the inner-loop stability augmentation.

The second type of failure will be detected quickly by the pilot; however, the outer loop limits serve to provide time for the pilot to apply corrective actions.

F-4 Simulation

In contrast to the C-141 data, in the F-4 simulation where failures were inserted into the system without the pilot's knowledge, because the failure detection system replaced the failed channel with an operating channel and because there were no transients in the system, the pilot was not even aware of any failure, except for noting a light on the flight panel, indicating that a channel had failed.

The point is that the F-4 was quad-redundant and a true fly-by-wire flight control system, and hence demonstrated the reliability in event of failure. The C-141 was a hybrid system and if a failure occurred in a nonredundant portion, the pilot might be required to respond rather quickly to overcome an undesired maneuver or the resultant unknown consequences.

One can conclude that for future aircraft it is probably far better to add control channel redundancy and electronics than to use hybrid types that could be more economical but surely have less reliability and safety.

Afterword

Both of the aircraft, i.e., F-4 #12200 and the B-47 used on the fly-by-wire research and development programs, are now retired and on display at the Air Force Museum, Wright-Patterson Air Force Base, Ohio.

Addressess of the persons who produced this document are:

Vernon R. Schmitt
3970 Alkire Road
Grove City, OH 43123

James W. Morris
2619 Bethany Court
Dayton, OH 45415

Gavin D. Jenney
7060 Cliftwood Place
Dayton, OH 45424

Index

About the Authors

Vernon R. Schmitt

Vernon R. Schmitt, a native of Columbus, Ohio, is an alumnus of the Ohio State University College of Engineering. He retired as a civilian project engineer from the U.S. Air Force Flight Dynamics Laboratory at Wright-Patterson Air Force Base after spending more than 30 years on research and development programs related to flight control and servo actuation systems. After retiring, Mr. Schmitt was a consultant to the Flight Dynamics Laboratory for the next 15 years in support of its flight control research programs. He was responsible for research and development efforts on fly-by-wire by the Douglas Aircraft Company, Sperry Phoenix Corporation, LTV E-Systems, and the in-house work on the B-47 aircraft. Following World War II military service, Mr. Schmitt spent six years in the Guided Missile Section at Wright-Patterson Air Force Base and was responsible for the flight control systems on the Matador and Rascal missiles. Part of his five-year military career was spent with the Special Weapons Group, which dealt with controlled bombs, now referred to as "smart bombs," and as instructor on bombsights and the C-1 autopilot.

James W. Morris

James W. Morris earned a B.S. in Mechanical Engineering at the University of Alabama and an M.S. in Armament Engineering at the U.S. Air Force Institute of Technology. He has 43 years of combined experience with the U.S. Air Force as a military officer and a civilian consultant to U.S. Air Force research and development projects. Mr. Morris rose through the ranks as project engineer on a diverse collection of technical development programs including conventional aircraft armaments; guided aircraft rockets; fly-by-wire flight control systems; digital computers; satellite and space vehicle controls; and systems integration of flight control with fire control, propulsion, and navigation. At the U.S. Air Force Flight Dynamics Laboratory, Mr. Morris was Program Manager of the U.S. Air Force Survivable Flight Control System Program, a program that made fly-by-wire an operational reality. His efforts ensured successful transition of this technology to aircraft such as the F-16, F-18, B-1, B-2, F-20, F-22, and F-117, and for engine controls. For these accomplishments, Mr. Morris was awarded the U.S. Air Force Meritorious Service Award and the AFA Meritorious Award for Program Management. Mr. Morris now provides technical, management, and administrative consultant services to the Flight Control Division of the U.S. Air Force Research Laboratory.

Gavin D. Jenney

Gavin D. Jenney earned a B.S. in Mechanical Engineering at Lafayette College, an M.S. in Mechanical Engineering at the University of Rochester, a Ph.D. in Engineering at Ohio State University, and an M.B.A. at Wright State University. Since 1974, Dr. Jenney has been president of Dynamic Controls, Inc. and has directed the company's technical effort in research and development programs with the U.S. Air Force at Wright-Patterson Air Force Base, Ohio. These programs have been involved with advanced electrohydraulic control system technology. Prior to starting Dynamic Controls, Inc., Dr. Jenney was principal investigator for the Hydraulic Research and Manufacturing Company on four consecutive research and development programs at Wright-Patterson Air Force Base. As part of this effort, he directed the design of the two-axis B-47 fly-by-wire control system and flew with the test aircraft as test engineer. Dr. Jenney has written more than 25 publications on the results of research and development in flight control systems, and he holds eight patents in this area.